LECTURE NOTES ON FULLERENE CHEMISTRY

A Handbook for Chemists

LECTURE NOTES ON FULLERENE CHEMISTRY

A Handbook for Chemists

Roger Taylor

The Chemistry Laboratory, CPES School,
University of Sussex

Imperial College Press

Published by

Imperial College Press
203 Electrical Engineering Building
Imperial College
London SW7 2BT

Distributed by

World Scientific Publishing Co. Pte. Ltd.
P O Box 128, Farrer Road, Singapore 912805
USA office: Suite 1B, 1060 Main Street, River Edge, NJ 07661
UK office: 57 Shelton Street, Covent Garden, London WC2H 9HE

British Library Cataloguing-in-Publication Data
A catalogue record for this book is available from the British Library.

ISBN 1-86094-104-4
ISBN 1-86094-109-5 (pbk)

Printed in Singapore.

Contents

Preface   xiii

1 Introduction   **1**
  1.1 Occurrence   1
  1.2 The Carbon Allotropes   2
  1.3 Discovery of the Fullerenes   4
  1.4 Isolation and Characterisation of the Fullerenes   8
  1.5 Endohedral Fullerenes (*Incar*-fullerenes)   11
  1.6 Multiwall or Nested Fullerenes   12
  1.7 Opened Fullerenes   13
  1.8 Elongated Fullerenes — Nanotubes   13
  1.9 Heterofullerenes   14

2 The Structure and Properties of Fullerenes   **17**
  2.1 Symmetry and Schlegel Diagrams   17
  2.2 Higher Fullerenes   19
  2.3 Preparation of Fullerenes   19
  2.4 Nomenclature of Fullerenes   21
  2.5 *Quasi*-fullerenes   26
  2.6 Dimeric and Fused Fullerenes   26
  2.7 Properties of Fullerenes   27
    2.7.1 Bond Lengths   27
    2.7.2 Solubility   28
    2.7.3 Thermodynamic and Related Properties 29
    2.7.4 Oxidation and Reduction: Voltammetry 30
    2.7.5 Spectroscopic Properties   31
  2.8 The Structure and Properties of Nanotubes   32

3 Addition Patterns   **42**
  3.1 Introduction   42
  3.2 Addends with No 1,2-Eclipsing Interactions   45
  3.3 Addends with Small Eclipsing Interactions   47
  3.4 Addends with Larger Steric Interactions   49
  3.5 1,2- vs. 5,6-Addition to [70]Fullerene   52
  3.6 Addition to Higher Fullerenes   53

4 Hydrogenation   **56**
  4.1 Hydrogenation Conditions   56
    4.1.1 Hydrogenation with Hydrogen Under Pressure   56
    4.1.2 Hydrogenation with Hydrogen and a Catalyst   57
    4.1.3 Dissolving Metal Reductions   57
    4.1.4 Di-Imide Reduction   58
    4.1.5 Reduction Using Organometallic Reagents   58
    4.1.6 Transfer Hydrogenation   58
  4.2 Products of Hydrogenation   59
    4.2.1 Dihydrofullerenes   60
      (a) $C_{60}H_2$ 60
      (b) $C_{70}H_2$ 60
    4.2.2 Tetrahydrofullerenes   61
    4.2.3 $C_{60}H_6$ 63
    4.2.4 $C_{60}H_{18}$ 64
    4.2.5 $C_{60}H_{36}$ 64
    4.2.6 $C_{70}H_{8/10/12}$ 65
    4.2.7 $C_{70}H_{36/38/40}$ 66
    4.2.8 Hydrogenated Higher Fullerenes   66
    4.2.9 Formation of Methylene Adducts   66
  4.3 Theoretical Calculations   67
      (a) $C_{60}H_2$ and $C_{70}H_2$ 67
      (b) $C_{60}H_4$ 67
      (c) $C_{60}H_6$ 67
      (d) $C_{60}H_{36}$ 68

**5 Reduction by Electron Addition, and Reaction of
Fullerene Radical Anions with Electrophilies** **71**
  5.1 Formation of Radical Anions 71
  5.2 Reduction by Metals 73
  5.3 Reduction by Organic Donors 74
  5.4 Electrocrystallisation 74
  5.5 Electrochemical Reduction 75
  5.6 Reaction of Electrophiles with Radical Anions 75
  5.7 Reaction of Fullerenes with Alkali- and Alkaline
      Earth Metals 76

**6 Nucleophilic Addition, and Reaction of Fullerene
Anions with Electrophiles** **80**
  6.1 Reaction with Neutral Nucleophiles 80
  6.2 Anions Formed by Addition of a Negatively
      Charged Nucleophile 84
  6.3 Reactions of Anions with Electrophiles 85
  6.4 Anions Formed by Proton Loss 88

7 Radical Reactions **91**
  7.1 Halogenation 91
    7.1.1 Fluorination 92
        7.1.1.1 Fluorination with Fluorine Gas 92
        7.1.1.2 Fluorination with Metal Fluorides 97
        7.1.1.3 Fluorination with Krypton
                Difluoride 98
        7.1.1.4 Properties of Fluorofullerenes 99
    7.1.2 Clorination 99
    7.1.3 Bromination 101
    7.1.4 Iodination 103
  7.2 Reactions with Other Radicals 104
    7.2.1 Radical Reactions of [60]fullerene 104
    7.2.2 Radical Reactions of [70]fullerene 109

**8 Nucleophilic Substitution of Fullerenes: Fullerenes as
Electrophiles** **114**
  8.1 Methoxydehalogenation 115

8.2 Fullerenol Formation   115
    8.2.1   Hydroxydenitration and Hydroxydesulphonation   115
            8.2.1.1 Use of Aqueous HNO_3/H_2SO_4 116
            8.2.1.2 Use of Nitronium Tetrafluoroborate   117
            8.2.1.3 Use of Oleum   118
    8.2.2   Fullerenols from Nitrofullerenes   121
    8.2.3   Fullerenol Formation from
            Halogenofullerenes   121
    8.2.4   Fullerenol Formation via Hydroboration   122
    8.2.5   Fullerenol Formation Involving Reaction
            with Oxygen   122
8.3 Biological Applications of Fullerenols   122
8.4 Polymers Derived from Fullerenols   123
8.5 Allyldechlorination   123
8.6 Aryldehalogenation   123
    8.6.1   Substitutions in $C_{60}Cl_6$ 124
    8.6.2   Substitutions in $C_{70}Cl_{10}$ 129
    8.6.3   Substitutions in $C_{60}F_{18}$ 131
    8.6.4   Substitutions in Bromo[60]fullerenes   133
8.7 Fullerenylation   133

9 **Cycloadditions**   **137**
9.1 [1 + 2] Cycloadditions: Reactions That Produce
     Methano- and Homofullerenes and Their
     Homologues   137
    9.1.1   Addition of Carbon   138
            9.1.1.1 Additions Involving Carbenes,
                     and Azo Precursors   138
            9.1.1.2 Formation of Methanofullerenes
                     Through the Use of Ylides   147
            9.1.1.3 Formation of Methanofullerenes
                     by α-Halocarbanion Addition   147
            9.1.1.4 Miscellaneous Reactions Which
                     Produce Methanofullerenes   149
            9.1.1.5 Polyaddition and Tether-Directed
                     Reactions   150

	9.1.2	Addition of Nitrogen: Formation of Azahomofullerene and Epiminofullerenes	155
	9.1.3	Addition of Oxygen: Formation of Epoxides	160
	9.1.4	Addition of Silicon: Formation of Fullerene Siliranes	162
9.2	[2 + 2] Cycloadditions		163
	9.2.1	Additions by Alkenes	163
	9.2.2	Additions Involving Pseudo Alkenes (4-Membered Rings etc.)	165
	9.2.3	Addition of Benzyne	166
9.3	[3 + 2] Cycloadditions		167
	9.3.1	Formation of Cyclopentafullerenes	167
	9.3.2	Formation of Pyrrolidinofullerenes	169
	9.3.3	Formation of Pyrrolo[60]fullerenes	172
	9.3.4	Formation of Pyrazolo[60]fullerene	172
	9.3.5	Formation of Isoxazolofullerenes	173
	9.3.6	Formation of Oxazolo[60]fullerenes	174
	9.3.7	Formation of an Isothiazolo[60]fullerene	174
	9.3.8	Formation of a Thiazolo[60]fullerene	175
	9.3.9	Formation of Furano[60]fullerenes	175
	9.3.10	Formation of Disolanofullerenes	176
9.4	[4 + 2] Cycloadditions: Diels-Alder Reactions		176
	9.4.1	Reactions of *o*-Quinodimethane and Derivatives	177
	9.4.2	Reactions of Cyclobutane Derivatives	180
	9.4.3	Additions of Buta-1,3-Diene Derivatives	180
	9.4.4	Addition of Anthracene	183
	9.4.5	Addition of Cyclopentadiene, Furan and Derivatives	184
9.5	[6 + 2] Reactions		186
9.6	[8 + 2] Reactions		186

10 Oxidation and the Formation of Radical Cations and Cations — **200**

10.1 Addition of Oxygen — 200

10.2 Formation of Radical Cations   201
10.3 Formation of Cations   203

**11 Inorganic and Organometallic Derivatives of
Fullerenes**   **206**
11.1 Introduction   206
11.2 Osmylation   206
11.3 Fullerene Complexes of Iridium, Rhodium and
Cobalt   209
11.4 Fullerene Complexes of Platinum, Palladium and
Nickel   214
11.5 Fullerene Complexes of Iron and Ruthenium   216
11.6 Fullerene Complexes of Molybdenum, Vanadium
and Tantalum   217
11.7 Miscelaneous Fullerene-Metal Complexes   217

**12 Polymers, Dendrimers, Dimers, Dumb-bells and
Related Structures**   **223**
12.1 'Pearl-Necklace' Polymers   223
      12.1.1   Directly-Linked Fullerenes   223
      12.1.2   Indirectly-Linked Fullerenes   227
12.2 'Pendant-Chain' Polymers   229
      12.2.1   Attachment of a Fullerene to a Preformed
               Polymer Chain   230
      12.2.2   Polymers Formed from Addended Fullerenes   234
      12.2.3   Star Polymers   235
12.3 Electroactive Polymers   236
12.4 Fullerenes as Polymer Dopants   237
12.5 Dendrimers   238
12.6 Dimers   239
12.7 Dumb-Bells   240
12.8 Ball and Chain Structures   244

13 Heterofullerenes   **251**
13.1 Borafullerenes   252
13.2 Azafullerenes   252

14 The Chemistry of *Incar*-fullerene (Endohedral Fullerenes)   **257**

14.1 Nomenclature   257

14.2 Fullerenes with Incarcerated Metals   257

    14.2.1   Properties   259

    14.2.2   Chemistry   261

14.3 Fullerenes with Incarcerated Nitrogen   262

14.4 Fullerenes Having Incarcerated Noble Gases   262

Preface

It is just eight years ago that fullerenes became available for chemical study. During this time there has been a change in perception from one of initial excitement and indeed awe that fullerenes would undergo any kind of reaction, to one where these molecules are seen to conform to the long established rules of organic and mechanistic chemistry. Thus for example steric effects are seen to control the extent to which addition takes place, the relative positions of addends, the conformations of cycloadducts, and the ease of substitution of one addend by another. Electron-donating groups decrease the electron affinity and vice versa. Just as in polycyclic aromatic and heteroaromatic chemistry, minimal disturbance of the bond fixation in the ground states of the molecules is a critical factor in controlling the regiochemistry, so too this is important for fullerene reactions. One might assume therefore that the chemistry of fullerenes is now just routine, but this is certainly not the case. The understanding of the above factors is still far from the level where it is possible to predict reliably the outcome of many reactions. As examples, the location of addends during polyaddition is in most cases neither known nor accounted for, whilst the widely varying addition patterns obtained in additions to [70]fullerene remains unexplained. Unexpected and unexplained reactions crop up regularly, whilst the chemistry of higher fullerenes has hardly begun. Control of addition is also still in its infancy, and the scale of the problem is vastly greater than in the case of substitution of benzene and its derivatives, which even after 160 years of research, cannot always be positionally controlled as desired.

So whilst an impressive amount of information has been obtained in such a short time, due to the thousands of chemists world-wide who have enthusiastically set about the task, there is much to be done.

This book is aimed at a broad readership ranging from undergraduate to researchers at the cutting edge. Thus the initial chapters give an introduction. to fullerene discovery, and to fundamentals such as fullerene types, basic properties, nomenclature, etc., and the addition patterns insofar as they are presently understood. The subsequent chapters describe the main features of the known chemistry, divided into the main reaction areas. In order to aid readers seeking to find the latest information concerning a particular reaction, references to all the major work reported up to the end of 1997 and in some cases to mid-1998, are included. In order to limit the book length, the *Nature* format has been adopted whereby references to papers with more than six authors are described as e.g. 'Smith *et al.*', and hopefully no-one will be unduly offended by this. For the same reason, the number of fullerene diagrams, which are particularly space-demanding, has been restricted to a minimum.

Roger Taylor
August 1998

1

Introduction

1.1 Occurrence

Carbon comprises approximately 0.05% of the Earth's constituents, and is the fifteenth most abundant element. All life is carbon-based, and this gives rise to the description of the chemistry of carbon compounds as 'organic' (a term often misused these days). Our daily intake of carbon (in food) is approximately 300 g, while the average person contains 16 kg. Protein, DNA, chromosomes etc. are all made up from combinations of carbon-containing molecules. Air contains 0.35% of carbon dioxide, the gas that provides all of the carbon in our food and also is primarily responsible for keeping the Earth at a comfortable temperature.

Carbon was almost certainly the first element that man was aware of (though of course not recognised as such) since it is produced in the form of charcoal from burning wood. Charcoal, used in the earliest cave paintings and by artists today, consists of very small crystals of graphite, one of only two forms (allotropes) of carbon that until recently were known. The other form of carbon is diamond, also surely known from before recorded history. The slipperiness of graphite (used in pencil 'lead') and the hardness of diamond reflect their molecular structures, described below.

With these two forms of carbon having been known from time immemorial and the isolation of over 10,000,000 organic compounds, it seemed inconceivable that there might be, entirely unsuspected, a third form of carbon. One that is fundamentally different, is leading to an entirely new branch of organic chemistry and science, and is engaging the enthusiastic attention of thousands of scientists worldwide.

1.2 The Carbon Allotropes

Diamond is comprised of a three-dimensional network of sp^3-hybridised carbon atoms each of which is joined to four others by single (σ) bonds (Fig. 1.1). The angles between the bonds are 109° 28' which all point to the corner of a tetrahedron. Since the network has no defined edge, a molecule of diamond can in principle be infinitely large (though in practise the largest diamond found to date is only about 5 cm across). The finite size means that there has to be a boundary, at which there must be 'dangling' bonds. To remain like this confers some instability and so these bonds become attached to hydrogen atoms; the surface of any diamond is actually a mono-atomic layer of hydrogen! Diamond is thus a not quite pure form of carbon, though the hydrogen content of the overall structure is exceedingly small. The hardness

Fig. 1.1 Structure of diamond.

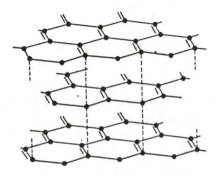

Fig. 1.2 Structure of graphite.

of diamond derives from the strong carbon-carbon bond network, with few structural discontinuities (and hence weaknesses).

Graphite is comprised of a network of sp^2-hybridised carbon atoms, each of which is joined to three others by means of two single (σ) bonds and one double ($\sigma + \pi$) bond (Fig. 1.2). The angles between the bonds here are 120° so that planar hexagonal arrays are formed. These arrays build up layers with relatively weak forces between them, and the layers can be sheared, one from another, which is what happens when you draw with a pencil. Like diamond, a molecule of graphite can in principle be infinitely large, but in practise is not, so here too there are dangling bonds at the boundaries, which have ultimately to be satisfied by bonding to hydrogen (so it too is not quite pure). The distance between the layers is approximately 3.35 Å (0.35 nm), and the relationship of one layer to another is such that alternate carbons in one layer lie over the centres of the hexagons in the adjacent one. (There are two ways of doing this, giving rise to α- and β-forms, but the difference does not concern us here.) The alternating double and single bonds within a given layer means that electrons can travel freely through the sheet, so making graphite an excellent conductor of electricity.

Even if the edge valencies of a graphite molecule are not satisfied, energetically this will not be too unfavourable since the fraction of atoms at the boundary is so very small. Consider however a small molecule built up from sp^2-hybridised carbon atoms, but which does not remain in the synthesis zone long enough to become very large, and is constructed in an atmosphere whereby no bonding of e.g. hydrogen to the boundary atoms is possible. Its small size means that the proportion of dangling bonds will be large and it will be very unstable. It can either combine with similar fragments to give a product with a reduced proportion of dangling bonds (which will be more stable), or it can curl up on itself so that the dangling bonds join, thereby giving a cage structure as found in fullerenes. The probability of this *intramolecular* combination of dangling bonds (giving fullerenes) is much higher than one involving *intermolecular* combinations (giving larger graphitic structures) because the collision frequency is higher, especially if the concentration of the fragments is low (as it will be in the vapour phase). Their formation is thus kinetically and not thermodynamically controlled, which is further confirmed by the greater abundance of the (less stable) C_{60} relative to C_{70}, since the collision frequency for formation of small structures is greater than for formation of larger ones. This is the key to the formation and structure of the fullerenes, the third form of carbon. Unlike the other two forms, fullerenes are molecules of *discrete size*, and their structure is a compromise between the desire of sp^2-hybridised carbon to exist in planar (and thus fully delocalised) arrays, and the need, during build up of these arrays, to quickly eliminate the dangling bonds. The resultant combination of strain (the strain energy makes up *ca.* 80% of the heat of formation)[1] and bond localisation makes fullerenes very reactive. Interestingly, the suggestion that graphite (containing discontinuities) could curl up to give a cage structure was humorously proposed by David Jones (under the pseudonym 'Daedalus') in 1966;[2] the discontinuities turned out to be pentagons, as described below.

1.3 Discovery of the Fullerenes

There have been many accounts of the discovery of fullerenes, some lengthy, others with significant omissions. A brief summary is given here because it is such an interesting example of both the scientific method and scientists at work.

The truncated icosahedron (which has 60 vertices and is the shape of C_{60}, the most abundant fullerene), has fascinated through the ages, due to the combination of twelve pentagons and twenty hexagons that produces the near spherical, and perfectly symmetrical structure; it is found in various parts of the world. Archimedes is credited with discovering the structure, Leonardo da Vinci included it in a book of his drawings, and it is found on a tomb in a Dorset church (Wimborne St. Giles), where it was thought to be a heraldic cabbage! (An interesting carbon connection here, for the German for cabbage is Kohl, whilst carbon is Kohlenstoff!) The same array is used for the 32-segmented bi-speed explosives surrounding the plutonium core of the British atomic bomb. (Designer Sir William Penney obtained the structure from his school geometry book, so it was clearly well-enough known.) And of course it is the structure of a football, which surprisingly played a significant part in the fullerene story. A larger similar looking spherical structure, but comprised of 12 pentagons and 50 hexagons (120 vertices) can be seen in the grounds of the Summer Palace in Beijing.

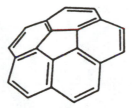

Fig. 1.3 Corannulene.

Fig. 1.4 First structural proposak of C_{60}, Japan.

Towards the end of the 1960's scientists were increasingly interested in non-planar aromatic structures, and shortly thereafter the saucer-shaped corannulene (Fig. 1.3) was synthesised.[3] In Japan, whilst playing football with his son, Eiji Osawa realised that a molecule made up of sp^2-hybridised carbons could have the football structure, a conjecture reinforced by the fact that the corannulene framework would fit precisely onto the surface. He therefore made the first proposal for C_{60} (Fig. 1.4) and suggested that it would be 'superaromatic'.[4] This prediction, also made subsequently by others, turned out to be incorrect for reasons to be described later. At this time also, the footballing enthusiasm of I. V. Stankevich of the topology group at the Nesmeyanov Institute in Moscow, caused this group independently to propose the C_{60} structure. Moreover the paper, published by Bochvar and Gal'pern in 1973 not only predicted some properties of C_{60}, but also of C_{20} (the smallest fullerene) as well,[5] and for good measure this group also built the first models (which still exist, Fig. 1.5) of both fullerenes. Theoretical calculations on C_{60} were also published in 1981 by Davidson.[6]

Fig. 1.5 Photographs of first models of C_{60} and C_{20}, Nesmeyanov Institute, Moscow.

The first spectroscopic evidence for C_{60} (and other fullerenes) was published in 1984 by Rohlfing, Cox and Kaldor from the Exxon research laboratories in New Jersey, USA.[7] They laser-vaporised graphite using a supersonic cluster beam apparatus (in which the species produced are analysed by a time-of-flight mass spectrometer), built to the design of Richard Smalley of Rice University, Texas. They found not only the carbon species C_{4n-1} which have maximum intensities for $n = 3, 4, 5, 6$, and thought to be linear chains (C_{11} was first observed by mass spectrometry in 1943 by Otto Hahn and coworkers),[8] but also species C_{2n}, $n \geq 10$ (Fig. 1.6). Moreover, these species showed a maximum intensity at C_{60} with C_{70} not far behind. The Exxon group were thus the first to obtain spectroscopic evidence for fullerenes, but did not attach any special significance to the C_{2n} species, nor draw any conclusions as to the structures.

A year later, Harold Kroto (of Sussex University, UK), being interested in the manner in which carbon chains might be produced by carbon stars, arranged to examine these chains by using the cluster beam apparatus of Richard Smalley and his group at Rice University, USA. They soon reproduced the earlier Exxon results, but notably, research student James Heath found conditions whereby C_{60} (Fig. 1.7) was formed *exclusively*, showing it to be a particularly stable species. After discounting various highly improbable structures, they concluded that the molecule must be a cage, and Smalley succeeded in building what has been described (but incorrectly — see Fig. 1.5) as the first model of C_{60}; like practically everyone else, the group were then unaware of the C_{60} proposals

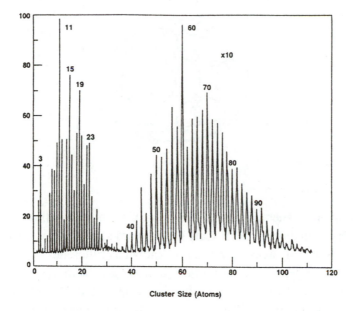

Fig. 1.6 First spectroscopic evidence for the existence of fullerenes.

Fig. 1.7 C_{60}.

that had been published fifteen years earlier. The paper describing this work was submitted to the journal Nature, on 12th September, 1985.[9] Heath also showed that lanthanum could be incorporated inside C_{60}, and this was the first formation of a member of the class of compounds known as *endohedral* fullerenes.[10] They named C_{60} as *Buckminsterfullerene*, because of the similarity of the structure to the geodesic structures widely credited to R. Buckminster Fuller (but in fact invented by Walter Bauersfeld of the Zeiss company for the

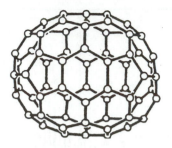

Fig. 1.8 C_{70}.

construction of a planetarium in 1922 in Jena, Germany). For this work, Kroto, Smalley, and collaborator R. F. Curl were awarded the Nobel prize in 1996. The general class of cages soon became known as fullerenes.

The next five years produced little progress apart from the proposal of the 'Isolated Pentagon Rule', i.e. that fullerenes would be stable only if they have non-adjacent pentagons.[11,12] The fact that C_{60} is the smallest fullerene for which this is true, and C_{70} (Fig. 1.8, the second most abundant fullerene in both the Exxon and Rice experiments) is the next smallest fullerene with non-adjacent pentagons, reinforced these structural explanations of the mass spectrometry observations.

The major breakthrough in fullerene science was made by Wolfgang Krätschmer of Heidelberg University and Donald Huffman of the University of Arizona. These astrophysicists were attempting to reproduce, by arc-vaporisation of graphite in a helium atmosphere, the formation of carbon grains thought to exist in space. Their experiments produced material that was benzene soluble (giving a red solution which deposited crystals on evaporation of the solvent), gave the predicted infra-red (IR) spectrum (consisting of only four bands) for C_{60}, whilst X-ray powder diffraction studies indicated a molecule of the predicted size for C_{60}. Their landmark paper[13] was submitted to *Nature* on 7th August, 1990.

1.4 Isolation and Characterisation of the Fullerenes

Krätschmer and Huffman had published a preliminary account of their IR results a few months earlier,[14] and at Sussex University, Kroto set student Jonathan

Hare the task of duplicating them. With help from Krätschmer concerning the experimental details (the helium pressure of *ca.* 70–100 mmHg is critical), Hare succeeded in producing a very small amount of soot which gave the reported IR spectrum, and found independently that it gave a red solution in benzene. It was apparent from the Krätschmer/Huffman paper (which had been sent to Sussex for refereeing) that the writers had not soxhlet-extracted their soot (they were unfamiliar with chemists' tools), had tried to purify their material by sublimation only (which does not work well for materials of similar structure), had not yet proved the fullerene structure, and there was no mass spectrum (one was provided after refereeing). Inevitably, this paper was going to trigger a mass scramble by research groups (some very well resourced) to duplicate (or even improve on) the results.

During discussion with Kroto, the writer offered to soxhlet extract Hare's soot (the most efficient way of separating soluble material from insoluble residues), and found that the 8 mg of fullerene extract so obtained would dissolve slightly in hexane, suggesting that it might be possible to separate the fullerenes by chromatography; being obviously non-polar molecules, chromatographic separation would be possible only by using a fairly non-polar solvent. Using a column of old (this was critical) neutral alumina, the writer found that a magenta band emerged on the column followed by a pinkish-red one.[15] These were C_{60} and C_{70} respectively (C_{70} solution is port-wine red on concentration); thus the beautiful colours of these fullerenes was discovered. This isolation of the pure materials was completed on 22nd August, 1990. Removal of the solvent gave mustard yellow, and red films for these fullerenes, respectively (much thicker films are both black). The 'old' alumina was critical because it contained traces of water, without which C_{70}, in particular, is degraded by the column.

Enough of the material was purified to run the mass-, UV- and IR spectra, and especially the ^{13}C NMR spectra; the latter gave a single line for C_{60} and five lines (in a ratio of 1:2:1:2:1 for the carbons *a,c,b,d,e*, respectively) for C_{70} (Fig. 1.9) which together unambiguously proved the conjectured structures. (C_{60} could have consisted of a 60-membered ring of sp-hybridised carbons atoms, and C_{70} a 70-membered ring, but both would then have given single-line spectra.) The NMR spectrum for C_{60}, perhaps the most notable one ever, is certainly the simplest! The paper describing this work, was sent to *Chemical Communications* on 7th September, 1990.[16]

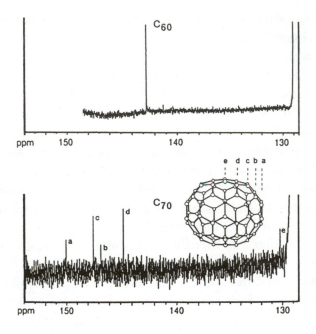

Fig. 1.9 NMR spectra of C_{60} and C_{70}.

During this frenetic period the writer discovered the initial properties of the fullerenes. First they tenaciously retained solvents (trapped in the lattice) as shown by the initial ^{13}C NMR spectrum which was largely that of benzene. Secondly, they oxidised rapidly (especially C_{60}), so that if the thin films were allowed to stay on the surface of the rotary evaporator flask for a few days, then part of the material could no longer be easily removed with benzene, and moreover acquired a pinkish tinge. (This rapid oxidation makes it highly improbable that fullerenes can occur naturally. One notable claim that they occur in Shungite, a carbon-rich deposit found near St. Petersburg, Russia,[17] has been disproved by mass spectrometry.[18]) Thirdly, pure material was much harder to dissolve (due to better packing in the lattice) than the impure precursor (which made it difficult to produce matrix solutions of the pure materials for FAB mass spectrometry). Fourthly, C_{70} was less soluble than C_{60}.

The explosive growth in fullerene research could now begin. The fullerene-containing soot was easy to produce, and the pure components could be isolated. Within a few weeks, improvements to the preparation and chromatographic purification had already appeared, the spectra had been duplicated (the assignments of the carbons for C_{70} were shown to be correct by 2-D analysis of ^{13}C enriched material),[19] and the first chemical reaction (reduction) was reported.[20] The initial work had shown the importance of the tools that are crucial to studies of the chemistry of fullerenes, namely mass spectrometry, high field Fourier transform NMR, FTIR, and chromatography. Soon added to these were the use of High Pressure Liquid Chromatography (HPLC),[21] and single crystal X-ray crystallography;[22] FT Raman spectroscopy[23] was also used. Fullerenes quickly became commercially available, and though the price for C_{60} was initially high (*ca.* $5,000/g), this was offset by studies being carried out on the *ca.* 10 mg scale; at the time of writing (1997), the price has fallen to *ca.* $50/g, and is available from suppliers in the UK, USA, Germany, China, and Russia.

1.5 Endohedral Fullerenes (*Incar*-fullerenes)

The formation by Heath of a fullerene with lanthanum incarcerated within was mentioned above. (The IUPAC names for these compounds are incarceranes, and the correct notation is $iLaC_{82}$, though $La@C_{82}$ etc. has entered into widespread use.) To date, the incarcerated elements have usually been rare earths or transition metals, and many compounds of this type have since been made through the use of graphite/metal (or metal oxide) rods in the arc-discharge process, or by laser vaporisation of similar materials. The yields are however, extremely small (typically 0.1% of the soot produced), though have been reported to be increased *ca.* tenfold by the use of metal carbides rather than metal oxides, thereby avoiding exposure of the endohedral fullerenes to oxygen (to which they are sensitive).[24]

Endohedral fullerenes were also reported (in a Soviet patent dated 1988) to be formed by addition of lanthanides to an iron/carbon eutechtic melt which was then quenched, the iron being removed subsequently with hydrochloric acid.[25] This work predated the more generally known macroscopic production of fullerenes noted above, and is notable by virtue of describing the solubility of fullerenes in toluene, and the fact that two scandium or yttrium atoms could be incarcerated, features that were confirmed only some years later. Surprisingly, it appears that no-one has attempted to repeat this work.

One very novel way of encapsulating radioactive atoms inside C_{82} involved first preparing $iGdC_{82}$ and then neutron irradiating it to transmute the most abundant gadolinium isotopes into ^{159}Gd and ^{161}Tb respectively.[26] These decay within a few days to ^{159}Tb and ^{161}Dy respectively; in due course applications may emerge for using these 'trapped' tracers.

Proof that the element is inside the cage rather than outside was provided in two ways: First it does not react with oxygen, ammonia, water etc., in contrast to the case when it is attached to the outside of the fullerene as in YC_{60}^+ species produced by ion-molecule reactions).[27] Secondly, fullerenes can be made to undergo photofragmentation involving C_2 loss which continues all the way down to C_{32}, the smallest viable empty fullerene that can be produced in this way without total disintegration of the cage structure. However, the rupture point is reached sooner with a metal inside, and at a point consistent with the size of the metal, e.g. at 36, 44 and 48 carbons respectively, for C_{60} containing lanthanum, potassium, and caesium, respectively.[28] This diagnostic tool is known as 'shrink-wrapping'. More recently, a synchrotron X-ray powder diffraction study has shown unambiguously that the metal is inside the cage (for iYC_{82}).[29]

At an early stage spectroscopic evidence indicated that it was possible to incarcerate more than one atom within a given cage, with cages comprised of 66 and 88 carbons being found to contain two and three lanthanum atoms, respectively.[28] Again, there are now many examples, and macroscopic quantities of endohedral fullerenes have recently been obtained, permitting examination of their properties and chemistry to commence (described in Chap. 14).

1.6 Multiwall or Nested Fullerenes

Even before the existence of fullerenes was conjectured, electron micrographs of some graphitised carbon particles obtained by Sumio Iijima showed the existence of nested fullerenes (though they were not then recognised as such).[30] The innermost cage had the dimensions predicted for C_{60}, and there were *ca.* 20 concentric cages. The size of the cages C_n that can be accommodated one within the other so as to leave acceptable intermolecular distances between them follow the series $n = m^2 60$, where m is an integer, i.e. 60, 240, 540, 960.... More recently, Daniel Ugarte showed that nested fullerenes (with up to

100 Å diameters) are formed from faceted carbon nanoparticles upon electron beam irradiation,[31] and these 'onions' can also be produced by laser melting of carbon under high pressure, by shock wave or plasma torch treatment of carbon soot, by the high temperature annealing of nanodiamonds, and by combustion of hydrocarbons.[32,33] However, investigation of the chemistry of these species, would seem to lie well into the future. An interesting development is that the distances between the layers is not constant, but reduces towards the centre of the nest. This creates a huge internal pressure to the extent that electron-beam irradiation of the cage at 700°C causes the inner layers to convert to diamond, thereby reducing the pressure.[34]

1.7 Opened Fullerenes

Fullerenes have been opened, as described in later chapters, and this had led to optimistic views concerning the possibility of incorporating small molecules within the cages, and even of resealing the opening. Neither of these objectives has yet been realised, and one problem not widely appreciated is that due to the ratio of internal to external volumes, at 3,000 atmospheres, only 1 molecule in 1,000 will enter the cage. Prospects for useful incorporation of molecules within the cages in this way, do not appear to be promising.

1.8 Elongated Fullerenes — Nanotubes

Another major development in fullerene science has been the discovery by Iijima of tube-like carbon structures, generally called nanotubes.[35] These are in effect elongated fullerenes, and just as the latter can be either empty or nested, so the nanotubes can be empty (referred to as single-wall nanotubes, abbreviated to SWNT) or nested (multiwall nanotubes, MWNT). They are commonly formed in the carbon deposit that collects at the cathode during arc-discharge (DC) of carbon rods,[35] and are also formed during hydrocarbon pyrolysis,[36] and combustion;[33] single-wall nanotubes are formed by arc discharge (DC) of carbon rods containing e.g. cobalt or iron/nickel.[37]

Pyrolytically-grown nanotubes are frequently spiralled, especially when these (finely divided) metals are used as catalysts.[38] Nanotubes are formed along with much other carbonaceous material, and these can be selectively removed by oxidation,[39] which, however, results in loss of 99% of the nanotubes; an alternative oxidation/hydrothermal procedure reduces this loss to 60%.[40]

There is considerable interest in the structure and properties of nanotubes (Chap. 2), because they promise to have applications as electrical, optical and mechanical devices; they also have promise as hydrogen storage materials, with consequent uses in fuel cells.[41] However, their main application will be as structural materials because of their high strength.

1.9 Heterofullerenes

Derivatives of fullerenes in which one or more carbon atoms are replaced by other elements are possible. Thus far fullerenes with one or two carbons replaced by boron (obtained by arc discharging between carbon rods containing boron nitride) have been observed spectroscopically,[42] whilst fullerenes containing nitrogen (azafullerenes) have not only been detected spectroscopically,[43] but obtained in macroscopic amounts by synthetic procedures. Because nitrogen has one more electron than carbon, the azafullerenes are radicals which can either dimerise to give e.g. $(C_{59}N)_2$[44] and $(C_{69}N)_2$[45] or add, for example, hydrogen to give $C_{59}NH$.[46] These compounds promise an extensive chemistry with additional features not available for fullerenes themselves because of the presence of the acidic hydrogen, and work undertaken thus far is described in Chap. 13.

References

1. R. C. Haddon, *Science*, **261** (1993) 1545.
2. D. E. H. Jones, *New Scientist*, **35** (1966) 245.
3. W. E. Barth and R. G. Lawton, *J. Am. Chem. Soc.*, **93** (1968) 1730.
4. E. Osawa, *Kagaku*, **25** (1970) 654; Z. Yoshida and E. Osawa, 'Aromaticity', Kagakudojin, Kyoto (1971), 174.
5. D. A. Bochvar and E. G. Gal'pern, *Proc. Acad. Sci., USSR*, **209** (1973) 239.
6. R. A. Davidson, *Theor. Chim. Acta*, **58** (1981) 193.
7. E. A. Rohlfing, D. M. Cox and A. Kaldor, *J. Chem. Phys.*, **81** (1984) 3322.

8. O. Hahn, F. Strassman, J. Mattauch and H. Ewald, *Z. Phys.*, **30** (1943) 598.

9. H. W. Kroto, J. R. Heath, S. C. O'Brien, R. F. Curl and R. E. Smalley, *Nature*, **318** (1985) 162.

10. J. R. Heath *et al.*, *J. Am. Chem. Soc.*, **107** (1985) 7779.

11. T. G. Schmalz, W. A. Seit, D. J. Klein and G. E. Hite, *J. Am. Chem. Soc.*, **110** (1988) 1113.

12. H. W. Kroto, *Nature*, **329** (1987) 529.

13. W. Krätschmer, L. D. Lamb, K. Fostiropoulos and D. Huffman, *Nature*, **347** (1990) 354.

14. W. Krätschmer, K. Fostiropoulos and D. Huffman, *Chem. Phys. Lett.*, **170** (1990) 167.

15. One report (J. Baggott, *Perfect Symmetry*) stated that three bands were observed. Another (H. Aldersley-Williams, *The Most Beautiful Molecule*) describes the pink band cluting before the magenta one. Both are incorrect.

16. R. Taylor, J. P. Hare, A. K. Abdul-Sada and H. W. Kroto, *J. Chem. Soc.*, *Chem. Commun.* (1990), 1423; this paper was published just 6 weeks after submission, probably a record for this journal. Thereafter, extremely rapid publication of results has become a hallmark of fullerene science.

17. P. R. Buseck, S. J. Tsipursky and R. Hettich, *Science*, **257** (1992) 215.

18. T. W. Ebbesen *et al.*, *Science*, **268** (1995) 1634; G. J. Langley, Y. Lyakhovetsky and R. Taylor, unpublished work.

19. R. D. Johnson, G. Meijer, J. R. Salem and D. S. Bethune, *J. Am. Chem. Soc.*, **113** (1991) 3619.

20. R. E. Haufler *et al.*, *J. Phys. Chem.*, **94** (1990) 8634.

21. J. P. Hare, H. W. Kroto and R. Taylor, *Chem. Phys. Lett.*, **177** (1991) 127.

22. J. M. Hawkins *et al.*, *J. Org. Chem.*, **55** (1990) 6250.

23. D. S. Bethune, G. Meijer, W. C. Tang and H. J. Rosen, *Chem. Phys. Lett.*, **174** (1990) 219; T. J. Dennis, J. P. Hare, H. W. Kroto, R. Taylor, D. R. M. Walton and P. J. Hendra, *Spectrochim. Acta*, **47A** (1991) 1289.

24. S. Bandow, H. Shinohara, Y. Saito, M. Ohkohchi and Y. Ando, *J. Phys. Chem.*, **97** (1993) 6101.

25. V. V. Levitskii and S. V. Domorov, *USSR Patent* (1988) SU1587000.

26. K. Kikuchi *et al.*, *J. Am. Chem. Soc.*, **116** (1994) 9775.

27. F. D. Weiss, J. L. Elkind, S. C. O'Brien, R. F. Curl and R. E. Smalley, *J. Am. Chem. Soc.*, **110** (1988) 4464.

28. Y. Chai *et al.*, *J. Phys. Chem.*, **95** (1991) 7564.

29. M. Takata *et al.*, *Nature*, **377** (1995) 46.

30. S. Iijima, *J. Cryst. Growth.*, **5** (1980) 675; *J. Phys. Chem.*, **91** (1987) 3466.

31. D. Ugarte, *Nature*, **359** (1992) 707.

32. K. Yamada, H. Kunishige and A. B. Sawaoka, *Naturwissenschaften*, **78** (1991) 450; N. Hatta and K. Murata, *Chem. Phys. Lett.*, **217** (1994) 398; V. L. Kuznetsov, A. L. Chuvilin, Y. V. Butenko, I. Y. Mal'kov and V. M. Titov, *Chem. Phys. Lett.*, **222** (1994) 343.

33. K. D. Chowdhury, J. B. Howard and J. B. Van der Sande, *J. Mater. Res.*, **11** (1996) 341.

34. F. Banhart and P. M. Ajayan, *Nature*, **382** (1996) 433.

35. S. Iijima, *Nature*, **354** (1991) 56.

36. C. J. Crowley and R. Taylor, unpublished work; M. Endo, K. Takeuchi, K. Kobori, K. Takahashi, H. W. Kroto and A. Sarkar, *Carbon*, **33** (1995) 873.

37. S. Iijima and T. Ichihashi, *Nature*, **363** (1993) 603; D. S. Bethune *et al.*, **363** (1993) 603; T. Guo, P. Nikolaev, A. Thess, D. T. Colbert and R. E. Smalley R. E., *Chem. Phys. Lett.*, **243** (1995) 49.

38. S. Amelinckx, X. B. Zhang, D. Bernaerts, X. F. Zhang, V. Ivanov and J. B. Nagy, *Science*, **265** (1994) 635; D. Bernaerts *et al.*, *Philos. Mag. A*, **71** (1995) 605.

39. T. W. Ebbesen, P. M. Ajayan, H. Hiura and K. Tanigaki, *Nature*, **367** (1994) 519.

40. K. Tohji *et al.*, *Nature*, **383** (1996) 679.

41. A. C. Dillon, K. R. Jones, T. A. Bekkedahl, C. H. Kiang, D. S. Bethune and M. J. Heben, *Nature*, **386** (1997) 377.

42. T. Guo, C. Jin and R. E. Smalley, *J. Phys. Chem.*, **95** (1991) 4948; V. Chai *et al.*, *J. Phys. Chem.*, **85** (1991) 7564; B. Cao, X. Zhou, Z. Shi, Z. Gu, H. Xiao and J. Wang, *Fullerene Sci. & Technol.*, **6** (1998) 639.

43. T. Pradeep, V. Vijayakrishnan, A. K. Santra and C. N. R. Rao, *J. Phys. Chem.*, **95** (1991) 10564; R. Yu *et al.*, *J. Phys. Chem.*, **99** (1995) 1818; J. Mattay *et al.*, *Tetrahedron*, **51** (1995) 6997.

44. J. C. Hummelen, B. Knight, J. Pavlovich, R. Gonzalez and F. Wudl, *Science*, **269** (1995) 1554.

45. B. Nuber and A. Hirsch, *Chem. Commun.*, (1996) 1421.

46. M. Keshavarz-K *et al.*, *Nature*, **383** (1996) 147.

2

The Structure and Properties of Fullerenes

Before describing fullerene chemistry, the structures and basic properties of the fullerenes must be considered. The properties of nanotubes (which as yet have no chemistry) are also considered here; the (brief) chemistry of *incar*-fullerenes is described in Chap. 14.

2.1 Symmetry and Schlegel Diagrams

For C_{60} there are 1812 possible structures,[1] but only one in which all the pentagons are non-adjacent, and this (having *icosahedral* symmetry, designated I_h) is thus the most stable isomer (and the only one isolated). Likewise there is only one such isomer (D_{5h} symmetry) for C_{70} (this possesses 5-fold symmetry axes, and a further plane of symmetry at right angles to these). Both these fullerenes are shown in Figs. 2.1 and 2.2 as *Schlegel* diagrams. These diagrams, which will be used extensively in this book, are 2-D representations of the 3-D structures; they are very suitable for demonstrating the chemistry, but rather less so for showing the symmetry. They are created by shrinking the polygons of the nearest face, expanding those of the far face, and turning the rear of the molecule inside out, and this is shown (Fig. 2.3) for C_{20}, the smallest fullerene that can in principle exist. Obviously each fullerene can have a large number of Schlegel diagrams, depending upon which of the polygons is chosen as the central view point, though in practice the polygon on a principal axis is normally used. A consequence of this is that Schlegel diagrams for C_{60} may have either a pentagon or hexagon at the centre.

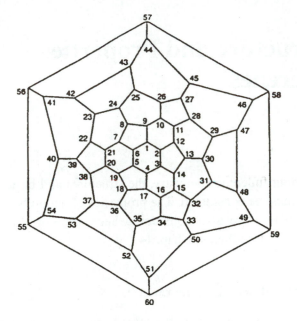

Fig. 2.1 Schlegel diagram for C$_{60}$.

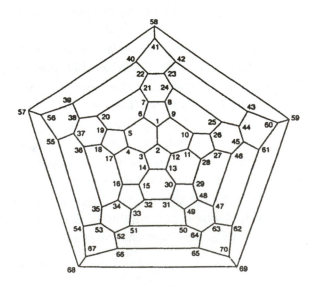

Fig. 2.2 Schlegel diagram for C$_{70}$.

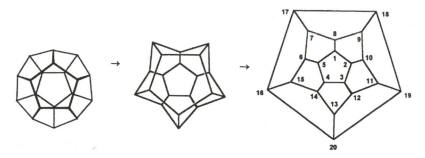

Fig. 2.3 Conversion of 3D structure for C_{20} into a Schlegel diagram.

2.2 Higher Fullerenes

Higher fullerene is the term applied to any fullerene possessing more than 70 carbon atoms. At the time of writing C_{72} and C_{74} have not been isolated, the latter because it is particularly insoluble. Not all fullerenes in the $C_{86}-ca$. C_{130} range have yet been separated, but obtaining them would be a straightforward (though exceedingly time-consuming and expensive) process (see Sec. 2.3). A further factor to be considered is the multiplicity of stable isomers that exist for fullerenes of 78 carbons or more. Thus even on separation of these fullerenes, it is difficult to make any mechanistic or quantitative evaluations based on the chemical derivatives, because these will consist of mixtures. A further separation of the *isomers* of the parent fullerenes is therefore required for rigorous analysis of the subsequent derivatives, which at present is barely practicable. For example, C_{84} consists of up to nine stable isomers.[2] The higher fullerenes formed in the largest quantities are C_{76},[3] C_{78},[4–6] and C_{84}.[5,6] C_{76} is elongated like C_{70} but flattened and twisted (it is chiral), the three main isomers of C_{78} are similarly elongated but have a slightly triangular cross section, whilst the two main isomers of C_{84} are almost spherical (one is chiral). At present the reported chemistry of these fullerenes is very limited.

2.3 Preparation of Fullerenes

Fullerenes can be made by either combustion[7] or pyrolysis[8] of aromatic hydrocarbons, but are most commonly and conveniently prepared by the arc-discharge process,[9] whereby carbon rods are vaporised in an atmosphere of *ca.* 100 Torr of either argon or helium (best) by passing a high electric current through them. Though sometimes referred to as resistive heating, this is incorrect because for success a small gap is required between the electrodes,[10] and this causes the tips to vaporise. Without this gap, the carbon rods would have to vaporise along their lengths (requiring a very high current), which is not observed. In general, rods of *ca.* 5 mm diameter are most suitable for currents of 100 A. Smaller rods give higher specific yields of fullerenes (but of course less overall yield per run); conversely larger diameter rods give lower specific yields. Larger reactor volumes give higher yields, since the fullerene containing soot is better able to diffuse away from the hot zone (and the intense UV light of the arc). Likewise yields are better if the inert gas is flowed through the arc zone. Either AC or DC may be used, the latter resulting in consumption of the anode only. This facilitates the use of anode rod autoloaders and continuous operation. Most workers report yields of fullerenes in their soot of *ca.* 10%, of which the respective yields of C_{60}, C_{70} and higher fullerenes (see below) are *ca.* 75, 24, and 1%.

Extraction (soxhlet) can most conveniently be performed with chloroform or toluene. The latter is more rapid, but there are risks of reaction brought about by the Lewes acid sites on the glass container; the problem is more severe if e.g. xylenes, mesitylene etc. are used since these are more susceptible to electrophilic substitution by the fullerene. Traces of these aromatics become trapped in the lattice and attempts to remove them by heating results in substantial reaction and degradation. This is not a problem for chemical use of the fullerenes, but is a significant one if accurate physical data are to be determined.

Purification is achieved by column chromatography. Initially alumina was used,[11] and subsequently, mixtures of alumina/charcoal with toluene/hexane elution.[12] The most satisfactory method now employs Elorit grade carbon granules with toluene elution, and in this way C_{60} can be obtained at a rate of 10 g/h; subsequent elution with 1,2-dichlorobenzene yields C_{70}.[13] Such have been the improvements to the preparation and purification processes, that C_{60} of a grade suitable for most chemical uses now costs only *ca.* $50/g. For higher fullerene separation, High Pressure Liquid Chromatography is used, the range of fullerene-dedicated Cosmosil columns with elution by either toluene, carbon disulphide or 1,2-dichlorobenzene, being very efficient for this purpose. Following purification it is necessary to wash samples with acetone to remove traces of hydrocarbons and especially di-octyl phthalate plasticiser that becomes concentrated from the solvents used. If traces of carbon disulphide are present in the soot extract prior to purification with alumina chromatography, this becomes converted to hydrogen sulphide which in turn is oxidised by the fullerene to sulphur, resulting in a yellow band eluting before C_{60}.[14]

2.4 Nomenclature of Fullerenes

Fullerenes are defined[15] as polyhedral closed cages made up entirely of n three-coordinate carbon atoms and having 12 pentagonal and $(n/2-10)$ hexagonal faces where $n \geq 20$ (but uniquely, not 22). The specific pentagon requirement is a consequence of the theorem of the Swiss mathematician Euler (1707–1783), which states that for any polygon the number of faces equals the sum of the edges minus vertices plus 2. For fullerenes the number of vertices is n, the number of edges (bonds) is $3n/2$ and hence the number of faces must be $n/2 + 2$. Since the number of hexagons is $n/2 - 10$, then the number of pentagons must be $(n/2 + 2) - (n/2 - 10) = 12$.

Fullerenes are three-dimensional annulenes and so the naming is similar to that for annulenes, C_{60} being thus described as [60]fullerene. As in the case of annulenes, the number in the square bracket indicates the number of atoms, and also the number of π-electrons, likewise double bonds; the overall number of bonds in $3n/2$. The symmetry of the fullerene should in principle be also included, e.g. [60-I_h]fullerene (where I_h describes the icosahedral symmetry) though for this fullerene it is generally omitted since it is clear which isomer being referred to. Likewise the symmetry for [70-D_{5h}]fullerene is generally omitted. In cases where more than one isomer can have the same symmetry, the isomers are distinguished by Roman numerals which refer to the ranking of the isomer in a ring spiral algorithm,[16] e.g. [78-C_{2v}(II)]fullerene etc.

The numbering for the different positions within a molecule is shown for the most common fullerenes in Figs. 2.1, 2.2, 2.4–2.9. The numbering is chosen so as to be *contiguous*, i.e. each carbon has a number that is $n + 1$ relative to a neighbour of number n. Groups attached to the cage are described as *addends*, not *substituents* (a common error), and the cage cannot be *substituted* unless a group (to be replaced) is already present on the cage. Because terms like bromo-, methyl- etc. are replacement terms, and since there is nothing to replace, the formal nomenclature procedure requires hydrogen to be 'added' to the molecule at the positions to be occupied by the addends. Thus for [60]fullerene with bromine added across the 1,2-positions, the correct name is 1,2-dibromo-1,2-dihydro[60]fullerene. This complicates matters somewhat and the hydro prefixes are sometimes omitted for simplicity. Note that in a particular reaction sequence, although given addends may occupy the same positions, their number locants may change as a result of adherence to the numbering rules.

Most 1,2-addition reactions take place across bonds between two hexagons (the so-called 6:6-bond), and if a second addition occurs across two such bonds, then eight different isomers are possible. These isomers can be described either by numbers or by a nomenclature introduced by Hirsch *et al.*, as shown in Fig. 2.10.[17] The positional numberings that are equivalent to this nomenclature are as follows: *cis*-1 (1,2,3,4); *cis*-2 (1,2,7,21); *cis*-3 (1,2,16,17); *e* (1,2,18,36); *trans*-1 (1,2,55,60); *trans*-2 (1,2,51,52); *trans*-3 (1,2,33,50); *trans*-4 (1,2,34,35).

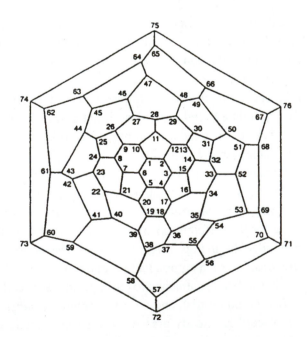

Fig. 2.4 Schlegel diagram for [76]fullerene.

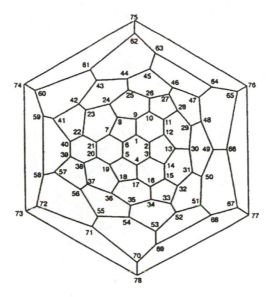

Fig. 2.5 Schlegel diagram for [78-C_{2v}(I)]fullerene.

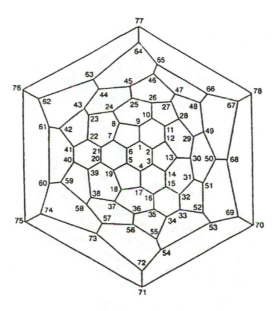

Fig. 2.6 Schlegel diagram for [78-C_{2v}(II)]fullerene.

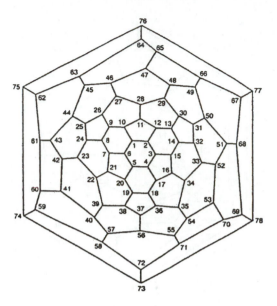

Fig. 2.7 Schlegel diagram for [78-D_3]fullerene.

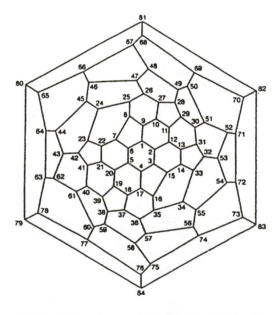

Fig. 2.8 Schlegel diagram for [84-D_2(IV)]fullerene.

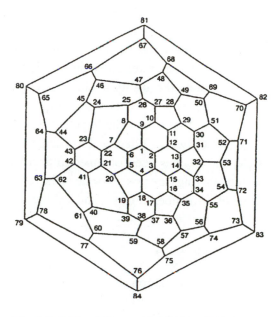

Fig. 2.9 Schlegel diagram for [84-D_{2d}(II)]fullerene.

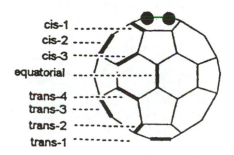

Fig. 2.10 Nomenclature system for describing the relationships of two addended bonds.

A further nomenclature system has been introduced for locating addends in a chiral fullerene (the chirality arising either from the parent molecule as in the case of [76-D_2]fullerene, or through the addition of an unsymmetrical addend to e.g. [70]fullerene) and is based upon the order in which the addends are encountered upon spiral numbering.[18] This system has not yet entered into general use, but will do so as more complicated structures are produced.

2.5 *Quasi*-fullerenes

Fullerenes containing other than just five- and six-membered rings are feasible. If a seven-membered ring is incorporated, then an additional five-membered ring will be required to maintain the ability to close the cage.[19] Such variants are known as *quasi*-fullerenes, and the sizes of the rings present are indicated in curved brackets in front of the name. The hypothetical octahedral C_{48}[20] is thus described as (4,6,8)[48-O_h]*quasi*-fullerene. Firm evidence for the existence of *quasi*-fullerenes C_{58} and C_{68} has been obtained in phenylated derivatives; in these, adjacent pentagonal rings are possible due to the relief of strain arising both from the presence of 7-membered rings, and sp^3-hybridised carbons.[21]

2.6 Dimeric and Fused Fullerenes

Compounds formed from a combination of fullerenes have now been isolated and characterised. These are $C_{120}O$ (Fig. 2.11),[22] $C_{120}O_2$ (Fig. 2.12),[23] C_{120} (Fig. 2.13)[24] and C_{119} (Fig. 2.14).[25] Spectroscopic evidence has also been obtained for C_{129} and C_{139} derived from C_{60}/C_{70} and C_{70}/C_{70} combinations.[26] An extensive chemistry for these compounds (and related ones yet to be isolated) may be anticipated.

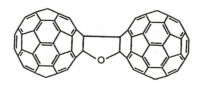

Fig. 2.11 $C_{120}O$.

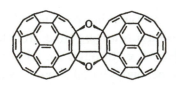

Fig. 2.12 $C_{120}O_2$.

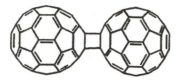

Fig. 2.13 C_{120}.

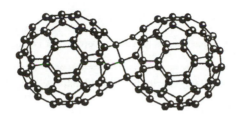

Fig. 2.14 C_{119}.

2.7 Properties of Fullerenes

The more important physical properties of fullerenes are given here, especially those of particular relevance to fullerene chemistry.

2.7.1 *Bond Lengths*

The bonds in [60]fullerene are of two different lengths. This is because it would increase strain to have a double bond in a pentagon.[27] Thus structure *a* in Fig. 2.15 is satisfactory in this respect, whereas *b* and *c* are not; this also partly explains why structures with adjacent pentagons are unstable. As a result, the lengths of bonds in a pentagon of [60]fullerene are *ca.* 1.45 Å, whereas those connecting two pentagons are *ca.* 1.40 Å.[28]

Likewise in [70]fullerene the bonds exocyclic to pentagons are shorter than those within pentagons. Table 2.1 gives the average bond lengths determined from neutron scattering[29] and X-ray measurements,[30] and π-bond orders calculated by the Hückel method.[31] The consequence of this bond localisation is that the fullerenes behave as alkenes rather than as aromatic compounds; a contributing factor also is the reduced overlap integral between adjacent carbons

a   *b*   *c*

Fig. 2.15 Favourable (*a*) and unfavourable (*b, c*) bond locations in pentagon/hexagon combinations.

Table 2.1 Bond lengths and π-bond orders for [70]fullerene.

Bond (see Fig. 19 for lettering)	Bond length/Å	π-bond orders
a–a	1.447	0.477
a–b	1.380	0.597
b–c	1.446	0.479
c–c	1.397	0.602
c–d	1.453	0.469
d–d	1.408	0.534
d–e	1.416	0.545
e–e	1.467	0.489

due to the curvature.[32] Thus their main reaction is *addition*. Nevertheless there is some aromaticity and this is an important factor in governing the direction of polyaddition, as described in the next chapter. The π-bond orders correlate almost linearly[33] with those calculated by the *ab initio* method,[34] and both correctly predict that the 1,2 (*a – b*) and 5,6 (*c – c*) bonds should be the most reactive. The Hückel calculations also correctly predict the positions of addition in [76]fullerene and [84]fullerene (including the isomer involved).

The cage-cage (centre) distances are 10.02 Å for [60]fullerene[35] and 10.61 Å for [70]fullerene.[36]

2.7.2 *Solubility*

The solubility of fullerenes decreases with increasing size. They are virtually insoluble in acetone, ethers, and alcohols. The solubility of [60]fullerene at room temperature in solvents commonly used are approximately (mg/ml): hexane, 0.04, dichloromethane, 0.26, carbon tetrachloride, 0.4, benzene, 1.7, toluene, 2.8, carbon disulphide, 6.5, 1,2-dichlorobenzene, 27, 1-methylnaphthalene, 33; for [70]fullerene the corresponding values are *ca.* 50% lower, though the actual value varies considerably with solvent. The solubility of [60]fullerene (and perhaps other fullerenes) decreases on lowering *and increasing* the temperature from around room temperature,[37] the extent of the change depending upon the extent to which the fullerene has been baked previously to release occluded solvents.[38] The existence of fullerene clusters, and which break up on raising the temperature, may be responsible.[39] Derivatives of fullerenes are generally more soluble than the parent molecule. Fullerenes react slowly with pyridine[40] (a solvent that is particularly suitable for extraction of *incar*-fullerenes) and with tetrahydrofuran.[41]

2.7.3 *Thermodynamic and Related Properties*

The heats of formation per C-atom for these fullerenes are not yet known with accuracy as different workers have obtained significantly different results, but are in the region of 10.0 and 9.3 kcal/mole respectively,[42] making them thermodynamically less stable than either diamond or graphite, the values for which are 0.4 and 0 kcal/mole respectively. Thus [70]fullerene is the more stable, and this is consistent with the expectation that the stability will increase as the surface becomes more planar, which it will do as the fullerenes become larger; for [60]fullerene some 80% of the heat of formation is due to the strain energy. The heats of sublimation increase with increasing fullerene size, and vary somewhat according to the method of determination. Recent values are 43.3, 47.8 and 50.1 kcal/mole for [60]-, [70]- and [84]fullerenes respectively;[43] these differences are too small to allow efficient separation of fullerenes by sublimation.

The electron affinities increase fairly regularly from values of 2.65 and 2.73 eV for [60]- and [70]fullerenes to 3.39 eV for [106]fullerene, though there is a sharp discontinuity for [72]- and [74]fullerenes, the values for which are 3.09 and 3.28 eV, respectively.[44] This may contribute to the failure to extract these fullerenes from fullerene-containing soot; [74]fullerene is known to be very insoluble in toluene. The electron affinities are increased by the presence on the cage of electron-withdrawing groups, e.g. the value for $C_{60}F_{48}$ is 4.06 eV.[45] The high electron affinities render the fullerenes susceptible to reaction with nucleophiles; the higher value for [70]fullerene compared with [60]fullerene leads to the expectation that it will be the more reactive. In solution at least this is usually not the case, and solvation may be at least partly responsible. This conjecture is supported by the electron affinities determined in hexane, whence the difference between the values for [60]- and [70]fullerene is reduced to 0.025 eV.[46]

At room temperature [60]fullerene exists as a face-centred cubic structure (fcc, lattice constant 14.17 Å),[47] the mass density being 1.72 gm/cm.[35] Below 260 K a phase transition to a simple cubic (sc) structure occurs.[48] At room temperature, the molecules in the lattice rotate (with ratcheting rather than constant angular momentum) at a rate of *ca.* $1 \times 10^{10}\,s^{-1}$,[49] which limits single crystal X-ray structure determinations either to fullerenes bearing heavy addends or to those having solvent interactions which slow rotation sufficiently.

2.7.4 *Oxidation and Reduction: Voltammetry*

The variation in electron affinity is paralleled by the LUMO energies.[44] The HOMO-LUMO gaps are 1.68 and 1.76 eV for [60]- and [70]fullerenes, respectively,[50] which is manifested in their stabilities. Both [60]- and [70]fullerenes have low-lying LUMO's and are therefore readily reduced and can acquire up to six electrons under electrochemical conditions;[51] other fullerenes are also readily reduced. It becomes increasingly difficult to add each subsequent electron, the electrochemical potentials for each addition becoming successively more negative. The presence of electron supplying groups e.g. alkyl on the cage causes the reduction potentials likewise to become more negative. In C_{60}^- the bond lengths are all the same *cf.* cyclo-octatetraene^{2-}.

The ease of reduction makes [60]fullerene a good oxidising agent and for example, it will oxidise H_2S to sulphur.[14] It is also a potent producer of singlet oxygen[52] and care should be exercised in handling it. Both [60]- and [70]fullerenes have low-lying HOMO levels, consequently electrochemical oxidation is difficult but achievable. Thus reversible one-electron oxidation waves (which give cation radicals with lifetimes > 30 s) have been obtained for each, with [70]fullerene undergoing a second one-electron oxidation;[53] addends appear generally to shift the oxidation potentials to less positive values. The oxidation potentials for [60]-, [70]- and [76]fullerenes have been determined as 1.26, 1.20 and 0.81 V, respectively.[54]

Fullerenes are unstable towards high temperature, and measurable decomposition can be observed for [60]fullerene above 750°C.[55] Decomposition is more rapid in the presence of oxygen (which [60]fullerene absorbs to the extent of 4 wt.-%).[56] Oxidation occurs at 250°C giving anhydrides, then acids, and eventually CO and CO_2.[56] Oxidative degradation in the presence of UV light is catalysed by both alumina[57] and titanium dioxide[58] and produces both water-soluble products[57] and epoxides.[58] The involvement of ozone, implicated here,[57] is confirmed by ozonolysis studies of fullerenes which lead to various oxide derivatives.[59]

2.7.5 Spectroscopic Properties

The m.o. energy levels govern the spectroscopic properties of the fullerenes. For example for [60]fullerene, four IR and ten Raman bands are observable, whilst for [70]fullerene these numbers become twelve and twenty-seven, respectively.[60] For [60]fullerene, the Raman line at 1469 cm^{-1} changes by 7 cm^{-1} for every unit charge on the cage, and also changes to 1457 and 1447 cm^{-1} if the cage becomes polymerised (by a 5 GPa pressure) to face-centred and rhombohedral structures, respectively;[61] the polymerisation is indicated by a reduction in the cage-cage (centre) distance to 9.22 Å, and by a greatly reduced solubility.

^{13}C NMR is the most widely used spectroscopic property of fullerenes. Due to the symmetries of the cages, the number of lines and their intensities for the common [X]fullerenes are: [60], 1×60; [70], 3×10, 2×20; [76], 19×4; $[C_{2v}(I)78]$ 3×2, 18×4; $[C_{2v}(II)78]$ 5×2, 17×4; $[D_3 78]$ 13×6; $[D_{2d}(II)84]$ 1×4, 10×8; $[D_2(IV)84]$ 21×4. Change in the symmetry upon addition to the cage is used to identify the positions of the addends. For example, addition of an unsymmetrical addend (e.g. $-CH_2O-$) across the 1,2-bond of [60]fullerene makes the symmetry C_s and the number of lines observed for the sp^2 carbons is then 13×4, 2×2, 2×1, with 2×1 for the sp^3 carbons.

The UV spectrum for [60]fullerene shows peaks at 213, 230(sh), 257(main), 329 and 406 nm (the values vary slightly with the solvent used) whilst that for [70]fullerene shows two main peaks at 214 and 236 nm (with a shoulder at 255 nm) and weaker peaks at 331, 360 and 378 nm. Weaker bands for both in the 420–700 nm region are due to forbidden singlet-singlet transitions and give rise to the magenta and red colours of [60]- and [70]fullerene, respectively.[10,11,62] UV spectra are used mainly to confirm the occurrence of 1,2-addition (i.e. across a 6:6 bond) to [60]fullerene, whence the products generally give a peak at *ca.* 435 nm,[63] whilst 1,4-addition produces a band at *ca.* 448 nm.[64] A water-soluble [60]fullerene carboxylic acid has been found to have cytoxicity and selective DNA-cleaving ability, but only upon irradiation by visible light, suggesting possible applications as photosensitive biochemical probes.[65]

2.8 The Structure and Properties of Nanotubes

Nanotubes are elongated fullerenes, the tubes being constructed entirely (except in certain circumstances) of hexagons, with six pentagons in the end caps in order that closure may occur (see Euler's theorem above). The pentagons can be arranged in a variety of ways giving different shapes to the caps (Fig. 2.16), and notably the shapes of the caps of the outer tubes parallel those of the inner ones; assembly evidently involves some template interactions between the layers (the distance between which is 0.34 nm). The tubes can also change diameter along their length, due to incorporation of pentagons and heptagons (Fig. 2.17).

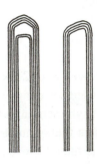

Fig. 2.16 Typical nanotube end caps.

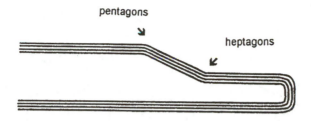

Fig. 2.17 Diameter change in nanotubes due to incorporation of heptagons.

Nanotubes can in principle have three different structure types, described as either [*m*,0]-, [*m*,*m*]-, or [*m*,*n*]-nanotubes (*m* and *n* are integers). Examples of their respective appearances are shown in Figs. 2.18(a–c), and they can be formed by rolling up a graphene sheet of hexagons in three different ways. This can be seen by reference to Fig. 2.19.

(a)

(b)

(c)

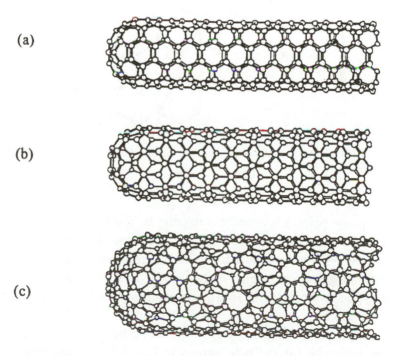

Fig. 2.18 Appearance of nanotubes of types: (a) *m*,*m*; (b) *m*,0; (c) *m*,*n*.

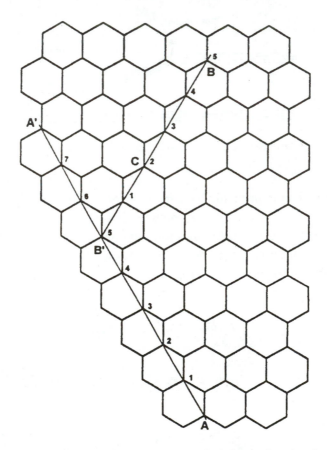

Fig. 2.19 Graphene sheet showing the different connections that produce, in principle, the three types of nanotubes.

   (i) Suppose point A is connected to point B. The locant of B is described in terms of two vectors of zig-zag carbon paths. One vector is the line A-A′ and the other is the line B-B′. Point B is reached from A by traversing 5-carbons along A-A′ and then 5-carbons along B-B′. The resultant nanotube (of the type shown in Fig. 2.18(a), and referred to as 'armchair') is thus described as [5,5].

   (ii) Suppose A is connected to B′. Only one vector is involved here, so the resultant nanotube (of the type shown in Fig. 2.18(b), and described as 'zig-zag') is described as [5,0].

(iii)   If A is connected to C, the nanotube (of the type shown in Fig. 2.18(c) is described as [5,2]. In this case the hexagons will spiral along the axis of the nanotube, which will be chiral.

Nanotubes are predicted to be either metallic or semiconducting according to their structures. Thus all $[m,m]$tubes and all $[3m + x, x]$ tubes (where $m = 1,2,3...; x = 0,1,2,3...$) are predicted to be metallic, and all others should be semiconducting. These predictions appears to be confirmed by measurements of the electrical conductivity (of multiwall tubes) which indicate that they have both metallic and non-metallic behaviour, with resistances varying between 2×10^2 and 5×10^8 ohms/μm.[66] They are further confirmed by the results of scanning tunnelling microscopy/ spectroscopy which show that both metallic and semi-conducting tubes are present, that these properties depend upon the wrapping angle, and that the bandgaps for both types are consistent with the theoretical predictions.[67] A particularly interesting nanotube would be one which was metallic at one end and semi-conducting at the other (which could be achieved by incorporating a seven-membered ring). This would then behave like an electronic switch.[68] The resistivity of single-wall nanotubes is decreased by up to 30-fold by doping with either potassium or bromine, due probably, to the increase in free charge carriers.[69]

Whereas multiwall nanotubes are largely straight, single-wall nanotubes are more flexible and can have curved arcs with curvature radii as low as 20 nm. Multiwall tubes have diameters between 3–30 nm, single tubes have diameters of 1–2 nm, and tube lengths can be up to 1000 nm. Nanotubes are incredibly stiff, and by measuring variations of tube vibrations with temperature, their Young's modulus has been found to lie between 0.5 and 4.2 Tpa[70,71] (the predicted theoretical maximum for single wall tubes is *ca.* 5 TPa); the tensile strength is 400 GPa compared to 20 GPa for steel. Formation of [10,10] single wall tubes (1.38 nm diameter) appears to be especially favourable (their caps correspond to one half of C_{240}, the next largest fullerene after C_{60} having icosahedral symmetry). They undergo tensile failure only at 40% elongation, are predicted to be metallic,[71] combine together into ropes, and the [10,10] structure has been confirmed by electron diffraction studies.[72] Nanotubes that have been opened by laser evaporation or by oxidative etching, show dramatically enhanced field emission of electrons, attributed to carbon atoms at the end of the tube being pulled out in an sp-hybridised chain, akin to the unravelling of knitting; such structures could provide atomic-scale field emitters.[73]

The most readily formed nanotubes are of the [m,m] type ('armchair'), and [10,10] in particular; a reason for the ease of formation of this type may be that the cage is readily extended by continual [4 + 2] cycloaddition. Nanotubes, consisting of mainly of the [10,10] 'armchair' type undergo coalescence when heated to 1500°C under vacuum, resulting in diameter doubling, and even trebling, the process being driven by a reduction in strain energy.[74]

Structural characterisation of any chemical derivatives of nanotubes by existing techniques will inevitably be near impossible, since the reactions will largely, if not entirely, take place in the pentagon regions. These constitute only a mere fraction of a percentage of the overall structure, hence spectroscopic analysis of the reaction products will be unachievable with existing techniques and instrumentation. Moreover the nanotubes are insoluble and this will be true of their reaction products, further adding to the problem.

So far the only reaction attempted is oxidation, and this results in the ends of the tubes being opened in an undefined way. Such opening accompanies oxidative removal of amorphous carbon material from nanotubes; the outer layers of multiwall nanotubes also get stripped away under these conditions. Opening can be achieved with either oxygen,[75] carbon dioxide,[76] or simplest of all, by merely boiling with nitric acid.[77] The opened tubes have then been filled (mostly only partially) by capillary action with elements such as lead,[78] bismuth[75] and selenium,[79] and by reaction with solutions of nitrates which, after calcining, leave a deposit of the metal oxide such as nickel oxide.[77] These oxides in turn can be reduced to the corresponding metal (e.g. Fe, Co and Ni)[77] thereby opening up the prospect of the manufacture of nanomagnets. Nanotubes with longer continuous fillings of Se, S, Sb and Ge have been prepared using the arc discharge method with graphite rods containing the corresponding elements.[80] The small proteins Zn_2Cd_5-metallothioein, cytochrome c_3 and β-lactamase I have also been immobilised in nanotubes, and retained a significant amount of catalytic activity indicating that no drastic conformational change had occurred;[81] this retention of activity is surprising considering the limited access to the protein.

A possible route to solving one problem associated with the post-production filling of nanotubes, namely the contamination with material attached to the outside of the tubes has been described. Tubes filled intially with rhodium chloride were washed with a reversed micelle mixture in non-polar solvents, and then the residual internal rhodium chloride was reduced to the metal by hydrogen.[82] On the other hand, material deposited on the outside of tubes has, in the case of ruthenium, been shown to catalyse the hydrogenation of cinnamaldehyde to the alcohol with much higher selectivity than use of a Rh/Al_2O_3 catalyst.[83]

References

1. D. E. Manolopolous, *Chem. Phys. Lett.*, **192** (1992) 330.
2. A. G. Avent, D. Dubois, A. Penicáud and R. Taylor, *J. Chem. Soc., Perkin Trans. 2*, 1997, 1907; M. Saunders *et al.*, *J. Am. Chem. Soc.*, **117** (1995) 9305; T. J. S. Dennis, personal communication.
3. R. Ettl, I. Chao, F. Diederich and R. L. Whetten, *Nature*, **353** (1991) 149.
4. F. Diederich, R. L. Whetten, C. Thilgen, R. Ettl, I. Chao and M. M. Alvarez, *Science*, **254** (1991) 73.
5. K. Kikuchi *et al.*, *Nature*, **357** (1992) 76.
6. R. Taylor, G. J. Langley, T. J. S. Dennis, H. W. Kroto and D. R. M. Walton, *J. Chem. Soc., Chem. Commun.*, (1992) 1043; R. Taylor, G. J. Langley, A. G. Avent, T. J. S. Dennis, H. W. Kroto and D. R. M. Walton, *J. Chem. Soc., Perkin Trans. 2*, (1993) 1029.
7. J. B. Howard, J. T. McKinnon, Y. Makarovsky, A. L. Lafleur and M. E. Johnson, *Nature*, **366** (1993) 729; J. B. Howrad, A. L. Lafleur, Y. Makarovsky, S. Mitra, C. J. Poe and T. K. Yadav, *Carbon*, **30** (1992) 1183.
8. R. Taylor, G. J. Langley, H. W. Kroto and D. M. Walton, *Nature*, **366** (1993) 729; R. Taylor and G. J. Langley, *Proc. Electrochem. Soc.*, **94–14** (1994) 68.
9. W. Krätschmer, L. D. Lamb, K. Fostiropoulos and D. R. Huffman, *Nature*, **347** (1990) 354.
10. J. P. Hare, H. W. Kroto and R. Taylor, *Chem. Phys. Lett.*, **177** (1991) 394.
11. R. Taylor, J. P. Hare, A. K. Abdul-Sada and H. W. Kroto, *J. Chem. Soc., Chem. Commun.*, (1990) 1423.
12. W. A. Scrivens, P. V. Bedworth and J. M. Tour, *J. Am. Chem. Soc.*, **114** (1992) 7917; W. A. Scrivens and J. M. Tour, *J. Org. Chem.*, **57** (1992)

6922; L. Isaacs, A. Wehrsig and F. Diederich, *Helv. Chim. Acta*, **76** (1993) 1231.

13. A. D. Darwish, H. W. Kroto, R. Taylor and D. R. M. Walton, *J. Chem. Soc., Chem. Commun.*, (1994) 15.

14. A. D. Darwish, H. W. Kroto, R. Taylor and D. M. Walton, *Fullerene Sci. & Technol.*, **1** (1993) 571.

15. E. W. Godly and R. Taylor, *Pure Applied Chem.*, **69** (1997) 1411.

16. P. W. Fowler and D. E. Manolopolous, 'Atlas of Fullerenes' Clarendon Pres, Oxford, (1995).

17. A. Hirsch, I. Lamparth and H. R. Karfunkel, *Angew. Chem. Intl. Edn. Engl.*, **33** (1994) 437.

18. C. Thilgen, A. Herrmann and F. Diederich, *Helv. Chim. Acta*, **80** (1997) 183.

19. R. Taylor, *Interdisciplinary Science Reviews*, **17** (1992) 161.

20. B. I. Dunlap and R. Taylor, *J. Phys. Chem.*, **98** (1994) 11018.

21. A. D. Darwish, P. R. Birkett, G. J. Langley, H. W. Kroto, R. Taylor and D. R. M. Walton, *Fullerene Sci. & Technol.*, **5** (1997) 705.

22. S. Lebedkin, S. Ballenweg, J. Gross, R. Taylor and W. Krätschmer, *Tetrahedron Lett.*, (1995) 4971.

23. A. Gromov, S. Lebedkin, S. Ballenweg, A. G. Avent, R. Taylor and W. Krätschmer, *Chem. Commun.*, (1997) 209.

24. G.-W. Wang, K. Komatsu, Y. Murata and M. Shiro, *Nature*, **387** (1997) 583.

25. A. Gromov, S. Ballenweg, S. Giesa, S. Lebedkin, W. E. Hull and W. Krätschmer, *Chem. Phys. Lett.*, **267** (1997) 460.

26. S. W. McElvany, J. H. Callagan, M. M. Ross, L. D. Lamb and D. R. Huffman, *Science*, **260** (1993) 1632.

27. R. Taylor, *Tetrahedron Lett.*, (1971) 3134.

28. C. S. Yannoni, P. P. Bernier, D. S. Bethune, G. Meijer and J. R. Salem, *J. Am. Chem. Soc.*, **113** (1991) 437; W. I. F. David *et al.*, *Nature*, **353** (1991) 147; K. Hedberg *et al.*, *Science*, **254** (1991) 410.

29. A. V. Nikolaev, T. J. S. Dennis, K. Prassides and A. Soper, *Chem. Phys. Lett.*, **223** (1994) 143.

30. S. van Smaalen, V. Petricek, J. L. de Boer, M. Dusek, M. A. Verheijen and G. Meijer, *Chem. Phys. Lett.*, **223** (1994) 323.

31. R. Taylor, *J. Chem. Soc., Perkin Trans. 2*, (1993) 813.

32. R. C. Haddon, L. E. Brus and K. Raghavachari, *Chem. Phys. Lett.*, **131** (1986) 165.
33. R. Taylor, *Proc. Electrochem. Soc.*, **97–14** (1997) 281.
34. J. Baker, P. W. Fowler, P. Lazzeretti, M. Malagoli and R. Zanasi, *Chem. Phys. Lett.*, **184** (1991) 182.
35. P. W. Stephens *et al.*, *Nature*, **351** (1971) 632.
36. G. B. Vaughan *et al.*, *Science*, **254** (1991) 1350; M. A. Verheijen *et al.*, *Chem. Phys.*, **166** (1992) 287.
37. R. S. Ruoff, D. S. Tse, R. Malhotra and D. C. Lorents, *J. Phys. Chem.*, **97** (1993) 3379.
38. A. L. Smith, personal communication.
39. V. N. Bezmelnitsin, A. V. Eletskii and E. V. Stepanov, *J. Phys. Chem.*, **98** (1994) 6665.
40. R. Taylor, unpublished work.
41. P. R. Birkett, A. D. Darwish, H. W. Kroto, G. J. Langley, R. Taylor and D. R. M. Walton, *J. Chem. Soc., Perkin Trans. 2*, 1995, 511.
42. H.-D. Beckhaus *et al.*, *Angew. Chem. Intl. Edn. Engl.*, **33** (1994) 996; H. P. Diogo, *et al.*, *J. Chem. Soc., Faraday Trans.*, **89** (1993) 3541.
43. V. Piacente, G. Gigli, P. Scardala, A. Giustini and D. Ferro, *J. Phys. Chem.*, **99** (1995) 14052; V. Piacente, G. Gigli, P. Scardala, A. Giustini and G. Bardi, *J. Phys. Chem.*, **100** (1996) 9815; V. Piacente, C. Palchetti, G. Gigli and P. Scardala, *J. Phys. Chem.*, **101** (1997) 4303.
44. O. V. Boltalina, E. V. Dashkova and L. N. Sidorov, *Chem. Phys. Lett.*, **256** (1996) 253.
45. R. Hettich, C. Jin and R. N. Compton, *Int. J. Mass Spectr. Ion Proc.*, **138** (1994) 26.
46. M. E. Burba, S. K. Lim and A. C. Albrecht, *J. Phys. Chem.*, **99** (1995) 11839.
47. A. R. Kortan, *et al.*, *Nature*, **355** (1992) 529.
48. A. Dworkin, *et al.*, *Compt. Rend. Acad. Sci. Series II*, **312** (1991) 979; **313** (1991) 1017.
49. C. S. Yannoni, R. D. Johnson, G. Meijer, D. S. Bethune and J. R. Salem, *J. Phys. Chem.*, **95** (1991) 9; R. Tycko *et al.*, *J. Phys. Chem.*, **95** (1991) 518.
50. W. Andreoni, F. Gygi and M. Parrinello, *Chem. Phys. Lett.*, **189** (1992) 241.

51. Q. Xie, E. Perz-Cordero and L. Echegoyen, *J. Am. Chem. Soc.*, **114** (1992) 11004.

52. J. W. Arbogast and C. S. Foote, *J. Am. Chem. Soc.*, **113** (1991) 8886.

53. D. Dubois, K. M. Kadish, S. Flanagan and L. J. Wilson, *J. Am. Chem. Soc.*, **113** (1991) 7773; Q. Xie, F. Arias and L. Echegoyen, *J. Am. Chem. Soc.*, **115** (1993) 9818.

54. Y. Yang *et al.*, *J. Am. Chem. Soc.*, **117** (1995) 7801.

55. I. Gilmour, J. P. Hare, H. W. Kroto and R. Taylor, *Lunar Planet Sci. Conf. XXII*, (1991) 445; M. Gevaert and P. M. Kamat, *J. Chem. Soc.*, *Chem. Commun.*, (1992) 1470; H. S. Chen, A. R. Kortan, R. C. Haddon and D. A. Fleming, *J. Phys. Chem.*, **96** (1992) 1016.

56. T. Arai, Y. Murakami, H. Suematsu, K. Kikuchi, Y. Achiba and I. Ikemoto, *Solid State Comm.*, **84** (1992) 827.

57. R. Taylor, *et al.*, *Nature*, **351** (1991) 277.

58. M. Gevaert and P. N. Kamat, *J. Chem. Soc.*, *Chem. Commun.*, (1992) 1470.

59. R. Malhotra, S. Kumar and A. Satyam, *J. Chem. Soc.*, *Chem. Commun.*, (1994) 1339; D. Heyman and L. P. F. Chibante, *Recl. Trav. Chim. Pays-Bas*, **112** (1993) 531, 639.

60. J. P. Hare, *et al.*, *J. Chem. Soc.*, *Chem. Commun.*, (1991) 412; T. J. Dennis, J. P. Hare, H. W. Kroto, R. Taylor, D. Walton and P. J. Hendra, *Spectrochim. Acta*, **47** (1991) 1289.

61. Y. Iwasa *et al.*, *Science*, **264** (1994) 1570.

62. Z. Gasyna, *et al.*, *Chem. Phys. Lett.*, **183** (1991) 283; S. Leach, *et al.*, *Chem Phys.*, **160** (1992) 451.

63. A. Hirsch, T. Grösser, A. Siebe and A. Soi, *Chem. Ber.*, **126** (1993) 1061.

64. T. Kitagawa, T. Tanaka, Y. Tanaka and K. Komatsu, *J. Org. Chem.*, **60** (1995) 1490; R. González, F. Wudl, D. L. Pole, P. K. Sharma and J. Warkentin, *J. Org. Chem.*, **61** (1996) 5837.

65. H. Tokuyama, S. Yamago, E. Nakamura, T. Shiraki and Y. Sugiura, *J. Am. Chem. Soc.*, **115** (1993) 7918.

66. T. W. Ebbesen, H. J. Lezec, H. Hiura, J. W. Bennett, H. F. Ghaemi and T. Thio, *Nature*, **382** (1996) 54.

67. J. W. G. Wildöer, L. C. Venema, A. G. Rinzler, R. E. Smalley and C. Dekker, *Nature*, **391** (1998) 59; T. W. Odoma, J. Huang, P. Kim and C. M. Lieber, *Nature*, **391** (1998) 62.

68. See R. E. Service, *Science*, **271** (1996) 1232 for general details.
69. R. R. Lee, H. J. Kim, J. E. Fischer, A. Thess and R. E. Smalley, *Nature*, **388** (1997) 255.
70. M. M. Treacy, T. W. Ebbesen and J. M. Gibson, *Nature*, **381** (1996) 678.
71. A. Thess *et al.*, *Science*, **273** (1996) 483.
72. J. M. Cowley, P. Nikolaev, A. Thess and R. E. Smalley, *Chem. Phys. Lett.*, **265** (1997) 379.
73. A. G. Rinzler *et al.*, *Science*, **269** (1995) 1550.
74. P. Nikolaev, A. Thess, A. G. Rinzler, D. T. Colbert and R. E. Smalley, *Chem. Phys. Lett.*, **266** (1997) 422.
75. P. M. Ajayan, T. W. Ebbesen, T. Ichibashi, S. Iijima, K. Tanigaki and H. Huira, *Nature*, **362** (1993) 522.
76. S. C. Tsang, P. J. F. Harris and M. L. H. Green, *Nature*, **362** (1993) 520.
77. S. C. Tsang, Y. K. Chen, P. J. F. Harris and M. L. H. Green, *Nature*, **372** (1994) 159.
78. P. M. Ajayan and S. Iijima *Nature*, **362** (1993) 522.
79. E. Dujardin, T. W. Ebbesen, H. Hiura and K. Tanigaki, *Science*, **265** (1994) 1850.
80. A. Loiseau and H. Pascard, *Chem. Phys. Lett.*, **256** (1996) 246.
81. S. C. Tsang, J. J. Davis, M. L. H. Green, H. A. O. Hill, Y. C. Leung and P. J. Sadler, *J. Chem. Soc., Chem. Commun.*, (1995) 1803.
82. J. Cook, J. Sloan, J. R. Heesoms, J. Hammer and M. L. H. Green, *Chem. Commun.*, (1996) 2673.
83. J. M. Planeix *et al.*, *J. Am. Chem. Soc.*, **116** (1994) 7935.

3

Addition Patterns

3.1 Introduction

Before discussing the chemistry of the fullerenes, the various addition patterns that occur and the possible reasons for them, must be considered. Whilst the factors governing the addition of two or even four groups to fullerene are reasonably well understood, for higher addition levels much more work is needed both to ascertain the regiochemistry of addition (defined so far in only a few cases) and the underlying reasons for it. The problems are complicated by different reagents producing different addition patterns, as described below.

A reagent may add by insertion either into a 6:5 σ-bond ([1 + 2] cycloadditions), or into a 6:6 π-bond (usually referred to, and hereafter, as addition across a 6:6 bond). The latter, which is much more common, is generally diagnosed for example by the formation of a band at *ca.* 435 nm in the UV spectrum of the derivative (see Sec. 2.7.5). The main factors governing the 6:6 π-bond addition are as follows:

(i) The cages are relatively non-aromatic, and so any addition pattern that increases their aromaticity will increase stability and will therefore be favourable.

(ii) The rigidity of the cages is such that significant twisting to avoid eclipsing interactions, where these occur, is not possible. Consequently, steric effects are of considerable importance. The differences in the bond lengths in the molecules also affects the steric interactions. Steric effects also govern the choice between addition across a 6:6-bond (giving a methanofullerene) or insertion into a 6:5-bond giving a homofullerene (see Sec. 9.1).

(iii) Because the molecules have some aromaticity and hence electron delocalisation, addition across the 6:6-bond between a pair of adjacent hexagons increases the electron localisation in each of these rings. Therefore, *steric considerations permitting* (see below), further addition will tend to take place across one of the other double bonds in either of these hexagonal rings (there are four such bonds in total and in [60]fullerene they are all equivalent). One hexagon will therefore have addition across two of its double bonds, which should increase localisation further for the remaining double bond. This is observed for example in hydrogenation, and the additions of oxygen. Thereafter steric interactions intervene so that further addition takes place preferentially in the other hexagon, in which there is a choice of two positions. This gives rise to either a 'T' or 'S' addition pattern as shown in Fig. 3.1. These patterns are commonly observed in e.g. hydrogenation, fluorination, epoxide formation etc. There are greater eclipsing interactions in the 'T' structure compared to the 'S' structure, consequently, the latter is predicted by *ab initio* calculations to be the more stable.[1]

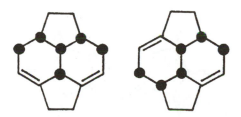

Fig. 3.1 'T' and 'S' addition patterns.

(iv) There is an unfavourable electron distribution around the equator of [70]fullerene. This can be very simply seen from a valence bond representation drawn so that double bonds are exocyclic to pentagons (Fig. 3.2). This results in non-alternation of the electrons in the five hexagons around the [70]fullerene equator or waist. 1,4-Addition giving *p*-quinonoid-type structures is therefore favoured here, as shown in Fig. 3.3, and is notable in chlorination (Chap. 7). 1,2-Addition giving an *o*-quinonoid structure can also occur (Fig. 3.4), even though this requires addition across the 7,8-bond of low π-bond order; this has been observed in the reaction with benzyne (Sec. 9.2).

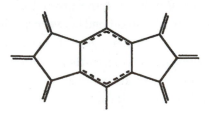

Fig. 3.2 Consequence in [70]fullerene of arranging all double bonds exocyclic to pentagons.

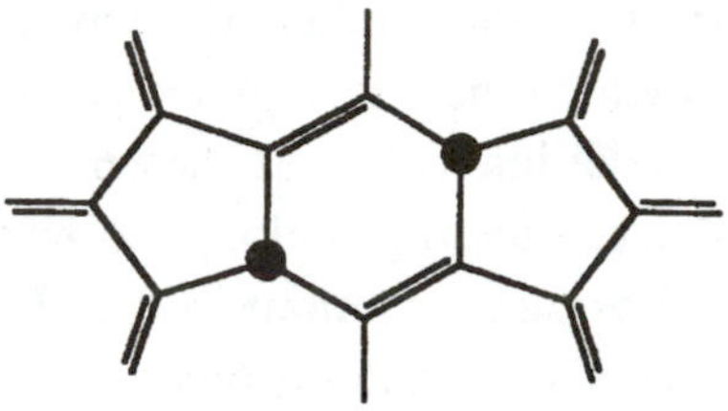

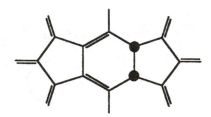

Fig. 3.3 1,4 eq. addition to [70]fullerene.

Fig. 3.4 1,2 eq. addition to [70]fullerene.

(v) Additions that place double bonds in pentagons will be unfavourable unless the strain in the pentagons is reduced through the presence of sp^3-hybridised carbons, or there are other stabilising features.

(vi) When steric effects favour other than a 1,2,3,4-addition pattern, seven other possibilities are possible for [60]fullerene (see Sec. 2.4). For [70]fullerene a similar pattern (in this case numbered 1,2,5,6) will produce chiral derivatives, whilst addition across a 1,2-bond at one end of the molecule, and an equivalent bond at the other, produces two chiral and one achiral possibilities.[2]

(vii) The relative proportions of addition across the 1,2- *vs.* the 5,6-bond of [70]fullerene varies greatly from reaction to reaction,[3] for reasons entirely unclear at present.

Three main addition patterns are currently evident:

3.2 Addends with No 1,2-Eclipsing Interactions

These are the cycloadditions which are categorised as [1 + 2], [2 + 2], [3 + 2] and [4 + 2] reactions, and in these a bridging group is joined to two adjacent carbon atoms. Typical groups involving a single bridging atom are CR_2 (R = H, Ph, CO_2Et etc.), NR, O and PtR_2 (R = Et, Ph); for examples of other cycloadditions see Chap. 9. These groups usually add across 6:6-bonds which are those having the highest bond order, (though one example is now known for [70]fullerene,[4] where as noted above, addition occurs across a 6,5-bond of low bond order because this creates a more favourable bond alternation in the product). When bulky groups are attached to the bridging atom in [1 + 2] additions and probably in all other cases, the 'T' and 'S' patterns described above are precluded. Instead up to six of these groups add to [60]fullerene, usually in an octahedral addition pattern (Fig. 3.5). This not only avoids steric interactions between the components of the addends, but creates eight hexagonal rings having increased delocalisation and aromaticity (Fig. 3.6), and is therefore favoured.[5] Increased delocalisation in these hexagons is possible due to reduction in strain in the adjacent pentagons (see Sec. 2.7) which can therefore accommodate some increased double bond character in the bond common to the pentagon and hexagon. This is confirmed by bond lengths in the hexagonal rings which are 1.39 and 1.42 Å thus showing less bond alternation than in [60]fullerene itself.[6] The octahedral array for addition to [60]fullerene has been structurally confirmed in the case of the addition of $Pt(PEt_3)_2$ (Fig. 3.5)[7] and $C(CO_2Et)_2$,[6] and indicated by mass spectrometry in the addition of CPh_2,[8] cyclopropylCCl,[9] benzyne,[10] cyclopentadiene,[11] morpholine[12] and methylene.[13]

In the octahedral pattern described above, the molecule can be considered as having two groups of triple addends grouped around polar benzenoid rings. At one 'pole' the addends are arranged clockwise, and at the other they are anticlockwise. However, an alternative (less stable) six-fold addition pattern has been observed recently,[14] in which both arrangements are either clockwise

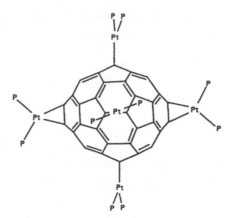

Fig. 3.5 Octahedral addition to [60]fullerene.

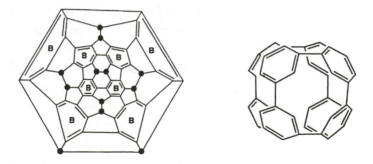

Fig. 3.6 Schlegel diagram of octahedral addition; eight aromatic (benzenoid) rings are created.

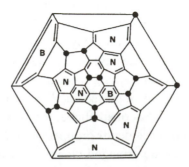

Fig. 3.7 Alternative six-fold addition to [60]fullerene; three naphthalenoid and two benzenoid aromatic patches are created.

or anticlockwise, one such being shown in Fig. 3.7, and this creates two benzenoid and three naphthalenoid patches. The various structures are thus diastereoisomeric.

For addition to [70]fullerene, if addition occurs across the 1,2-bond at one pole, then there is only the 31,32-bond in the same hemisphere (see Fig. 2.2) that gives the same relative positions of the addends as any pair in the octahedrally addended [60]fullerene. However, this does not lead on to a product of significantly increased aromaticity, and moreover, bonds in other hemisphere (41,58; 56,57; 67,68) have higher bond orders than the 31,32-bond. Accordingly addition preferentially occurs in the other hemisphere giving three structural products (two of which, as noted above, will be chiral) or five if the addends themselves are chiral.[15]

3.3 Addends with Small Eclipsing Interactions

Included here are the additions of hydrogen and fluorine (both are radical reactions), which give very similar patterns due to the small size of fluorine (its atomic radius is less than that of carbon because of the high electronegativity of fluorine). With [60]fullerene, each forms derivatives $C_{60}X_{18}$,[16] $C_{60}X_{36}$ and $C_{70}X_{36/38/40}$ (X = H, F).[17] $C_{60}F_{48}$ is also formed readily,[18] but $C_{60}H_{48}$ rather less so,[19] possibly because steric interactions at this addition level become more severe, the resultant energy barrier being more readily overcome by the lower activation energy for fluorination (which arises from the high C-F bond strength of the product). There is no evidence for further addition of hydrogen, and further addition of fluorine has been detected, but in only trace amounts.[20]

The small size of these addends means that 1,2-addition is possible, and further additions according to the 'S' and 'T' addition patterns noted above are particularly favourable (though others can occur, as described for hydrogenation in Chap. 4). Combinations of the 'S' and 'T' patterns lead ultimately for $C_{60}X_{18}$ to a crown structure in which a benzenoid ring is created (Fig. 3.8).[16] Further additions take place again according to these patterns to give a mixture of T and C_3 symmetry isomers (Figs. 3.9 and 3.10 respectively), each of which contains the $C_{60}X_{18}$ motif. Neither this motif nor either of the $C_{60}X_{36}$ motifs are evident in the structure of $C_{60}F_{48}$,[18] so that if these lower addended molecules are intermediates in the fluorination process, then at some stage, migration of fluorines across the cage must occur.

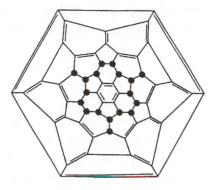

Fig. 3.8 Schlegel diagram of the addend location in $C_{60}X_{18}$ ($\bullet$ = H, F).

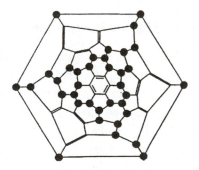

Fig. 3.9 T symmetry $C_{60}X_{36}$ ($\bullet$ = H, F).

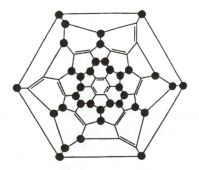

Fig. 3.10 C_3 symmetry $C_{60}X_{36}$ ($\bullet$ = H, F).

1,2-addition of hydrogen to [70]fullerene also occurs, but subsequent adjacent addition in the manner described for [60]fullerene in Fig. 3.1 leads to two possible tetra-adducts arising from addition across the 1,2,3,4- and 1,2,5,6-positions (see Sec. 4.2.2). The positions of the subsequent addition steps are not known, but lead ultimately to a mixture of $C_{70}X_{36/38/40}$ with the 38-addended species dominating. One explanation for this is that naphthalenoid and phenanthrenoid aromatic patches are created in the equatorial regions,[21] though more recent calculations suggest that the larger patches are unfavourable due to the surface curvature.[22] The lowest energy structures are those having many isolated pairs of addends, but it is uncertain as to whether or not these thermodynamically-favoured products will be preferred over the kinetic products resulting from the 'S' and 'T' addition patterns noted above.

The addition of oxygen also falls into this category, presumably because steric interactions between the *p*-orbitals are insufficient to prevent *bis* addition occurring within a given hexagon. Thus the oxygens in $C_{60}O_2$ are added across the 1,2- and 3,4-positions in a hexagon[23] and a third oxygen adds in the adjacent hexagon to give an arrangement analogous to the 'S' or 'T' patterns shown in Fig. 3.1.[24]

3.4 Addends with Larger Steric Interactions

This includes chlorine and bromine, aryl groups, and alkyl groups (though specific information regarding the latter is currently sparse). 1,2-addition gives rise to steric hindrance, but the alternative 1,4-addition places a double bond in a pentagon which is unfavourable energetically (see Fig. 3.10). For addends on intermediate size (such as small alkyl groups) the balance between 1,2-addition and 1,4-adddition will therefore depend upon the size of the steric interaction.

For larger groups, 1,2-addition of these larger groups is very sterically restricted, so that neither of the patterns noted in Secs. 3.2 and 3.3 are favourable. The trend here is for 1,4-addition to occur across hexagons (giving *p*-quinonoid structures), as can be seen from bromination of [60]fullerene whereby either six, eight, or twenty-four bromines can be added (Fig. 3.11). In the hexa-adduct, the addition of five groups provides a perfect sequence of

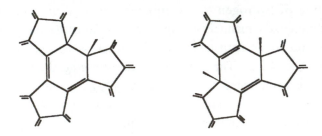

Fig. 3.10 1,2- vs. 1,4-addition.

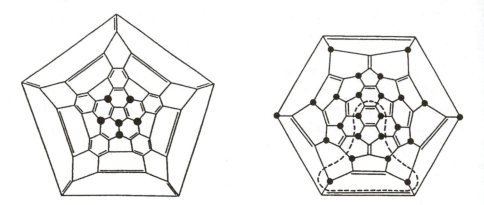

Fig. 3.11 The $C_{60}Br_6$,[25] $C_{60}Br_8$ (encircled)[27] and $C_{60}Br_{24}$[26] addition patterns (• = Br).

p-quinonoid rings (though this involves the unfavourable formation of a double bond in a pentagon), but leaves the central five-membered ring with an unpaired electron This can only be stabilised by the addition of a sixth bromine atom, so a sterically undesirable 1,2-addition becomes unavoidable. This sixth addition is not necessary in $C_{59}N$ (the central pentagon becomes 6π and therefore aromatic), and so only five groups add.[28]

Chlorination of [60]fullerene produces an identical pattern[26] as does reaction with alkyl radicals[29] *except* that in the latter case, the free electron can be stabilised by carbon-carbon hyperconjugation[30] and so addition (to $C_{60}R_5{}^{•}$) of a sixth alkyl group becomes unnecessary. The adjacency of two halogens creates a region of enhanced reactivity due to the bond lengthening arising from steric hindrance, and this is described in later chapters. In both $C_{60}Br_8$

and $C_{60}Br_{24}$ the pattern is dictated by the need to maintain a *p*-quinonoid pattern, yet avoid adjacencies; twenty-four addends are the most that can be attached to [60]fullerene with maintenance of this requirement (indeed there is no confirmed evidence for the attachment of any larger number of these halogens). Fig. 3.11 shows that the addend pattern in $C_{60}Br_8$ is a subset of that found in $C_{60}Br_{24}$ but the reason for the stability of the former subset is unclear, and may depend upon the derivative being rather insoluble, and so precipitates out of solution during bromination. One could anticipate that all derivatives $C_{60}X_{2n}$ where *n* ranges from 1–12 will be of comparable stability, the lack of their isolation merely representing the fact that addition cannot be controlled sufficiently. Indication that intermediate structures are possible comes from phenylation of [60]fullerene under appropriate conditions, which yields $C_{60}Ph_n$, where $n = 2,4,6,8,10,12$, though only the former compound has so far been structurally characterised.[31]

The 1,4-addition to [60]fullerene creates a problem that is overcome in $C_{60}X_6$ by placing two halogens on adjacent carbons. A related problem arises in chlorination (and probably bromination though it has not been examined) of [70]fullerene. It might be assumed that addition here would follow the pattern found in [60]fullerene (since the polar cap has the same structure as in [60]fullerene) thus giving e.g. $C_{70}Cl_6$ etc. However a more favourable alternative is available here. As noted in the introduction, bond localisation around the equatorial waist is such that 1,4-addition can occur here without introducing double bonds into pentagons (Fig. 3.2), and so ten chlorines are introduced in this region, and again two end up adjacent to each other.[32] As in the case of $C_{60}Cl_6$ this likewise creates a region of higher reactivity, which will be described in Chap. 8.

Addition of groups that are sterically more demanding than bromine or chlorine is less well documented, but there is a evidence that *ca.* 16–18 perfluoroalkyl groups can add to either [60] or [70]fullerenes,[33] and also 18 phenyl groups (Sec. 8.6.3) or 18 OH groups (Sec. 8.2.1.2) can add to [60]fullerene. Further work will be needed to elaborate on these data, and in particular to ascertain the positions of addition; it will be particularly interesting to see how (or indeed if) 18 groups are arranged over the cage surface in a symmetrical way because there appears to be no way in which this can be achieved.

3.5 1,2- vs. 5,6-Addition to [70]Fullerene

A problem that is as yet unresolved concerns addition across the 1,2- and 5,6-bonds in [70]fullerene. Although there is general consistency between reactions in that the 1,2-bond is always the more reactive, there is no obvious reason for the very wide variation in relative reactivities, seen from the collated data in Table 3.1. This is an area where further investigtion is necessary. The data for isoxazole additions shows that electronic factors must be significant.

Table 3.1

Addend	Isomer yields (%)				Ref.
	1,2	5,6	7,8	7,21	
Ir(CO)Cl(PPh$_3$)$_3$	100				34
H,Me H,Ph	100				35
C(CO$_2$Et)$_2$	100				15
(EtO)$_2$C = C(EtO)$_2$	96.5	3.5			36
H$_2$, (diimide)	96	4			37
Pyrazoline → CH$_2$	87	13			38
1,2,3,4,5-Pentamethylcyclopentadiene	mainly	detected			39
4-Methoxymethylisoxazole	75	25			40
OsO$_4$	68	22			41
Methylisoxazole	67	33			40
H$_2$, (diborane)	67	33			42
N-Methylidenemethylamine	46	41		13	43
Benzyne	43	34[a]	12[a]	12	44
(MeO)$_2$-orthoquinodimethane	24	10		1.5	45

[a]Updated positional assignments[44] following ^{13}C NMR analysis.[4]

3.6 Addition to Higher Fullerenes

This has been achieved so far in only a few reactions. Derivatives (including some polyadducts) have been isolated and either partly or fully characterised in osmylation of [76]fullerene (Sec. 11.2), cycloaddition of [76]- and [78]fullerenes (Secs. 9.1.1.3 and 9.4.1) and reaction of [84]fullerene with iridium (Sec. 11.3). In these reactions addition occurs across the positions calculated to have the highest π-densities.

Polyaddition of [76]-, [78]- and [84]fullerenes has been reported in the reaction with methylene groups (Sec. 4.2.9) and in hydrogenation (Sec. 4.2.8), the latter being accompanied by cage fragmentation to the lower fullerenes (this also happens in fluorination[45]). It is probable that polyaddition commences at the sites of highest π-density, further addition taking place according to the 'S' and 'T' patterns in Fig. 3.1.

References

1. P. A. Cahill, *Chem. Phys. Lett.*, **254** (1996) 257.
2. R. Taylor and D. R. M. Walton, *Nature*, **363** (1993) 685.
3. V. N. Bezmelnitsin, A. V. Eletskii, N. G. Schepetov, A. G. Avent and R. Taylor, *J. Chem. Soc., Perkin Trans. 2*, (1997) 683.
4. M. S. Meier, G. Wang, R. C. Haddon, C. P. Brock, M. A. Lloyd and J. P. Selegue, *J. Am. Chem. Soc.*, **120** (1998) 2337.
5. R. Taylor, *J. Chem. Soc., Perkin Trans. 2*, (1992) 1667; P. W. Fowler, D. J. Collins and S. J. Austin, *J. Chem. Soc., Perkin Trans. 2*, (1993) 275.
6. I. Lamparth, C. Maichle-Mössmer and A. Hirsch, *Angew. Chem. Intl. Edn. Engl.*, **34** (1995) 1667.
7. P. J. Fagan, J. C. Calbrese and B. Malone, *J. Am. Chem. Soc.*, **113** (1991) 9408.
8. T. Suzuki, K. C. Khemani, F. Wudl and O. Almarsson, *Science*, **254** (1991) 1186.
9. M. T. H. Liu, Y. N. Romashin, M. Kubota and M. Ohashi, *Organic Mass Spectrom.*, **29** (1994) 391.
10. M. Tsuda, T. Ishida, T. Nogami, S. Kurono and M. Ohashi, *Chem. Lett.*, (1992) 2333.
11. M. F. Meidine *et al.*, *J. Chem. Soc., Chem. Commun.*, (1993) 1342.

12. A. Hirsch, Q. Li and F. Wudl, *Angew. Chem. Intl. Edn. Engl.*, **30** (1991) 1309.

13. P. R. Birkett, A. D. Darwish, H. W. Kroto, G. J. Langley, R. Taylor and D. R. M. Walton, *J. Chem. Soc., Perkin Trans. 2*, (1995) 511.

14. Y. Rubin, personal communication.

15. C. Bingel and H. Schiffer, *Liebigs Ann. Chem.*, (1995) 1551; A. Herrmann, M. Rüttiman, C. Thilgen and F. Diederich, *Helv. Chim. Acta*, **78** (1995) 1673.

16. A. D. Darwish, A. G. Avent, R. Taylor and D. R. M. Walton, *J. Chem. Soc., Perkin Trans. 2*, (1996) 2051; O. V. Boltalina, V. Yu Markov, R. Taylor and M. P. Waugh, *Chem. Commun.*, (1996) 2549.

17. A. D. Darwish, A. K. Abdul-Sada, G. J. Langley, H. W. Kroto, R. Taylor and D. R. M. Walton, *J. Chem. Soc., Perkin Trans. 2*, (1995) 2359; O. V. Boltalina, A. Ya. Borschevskii, L. V. Sidorov, J. M. Street and R. Taylor, *J. Chem. Soc., Chem. Commun.*, (1996) 528.

18. A. A. Gakh, A. A. Tuinman, J. L. Adcock, R. A. Schleben and R. N. Compton, *J. Am. Chem. Soc.*, **116** (1994) 819; O. V. Boltalina, V. F. Bagryanstev, V. A. Seredenko, L. N. Sidorov, A. S. Zapolskii and R. Taylor, *J. Chem. Soc., Perkin Trans. 2*, (1996) 2275.

19. R. Loutfy, MER Corporation, personal communication.

20. H. Selig, K. Kniaz, G. B. M. Vaughan, J. E. Fischer and A. B. Smith, *Macromol. Symp.*, **82** (1994) 89; O. V. Boltalina, A. K. Abdul-Sada and R. Taylor, *J. Chem. Soc., Perkin Trans. 2*, (1995) 981.

21. R. Taylor, *J. Chem. Soc., Perkin Trans. 2*, (1994) 2497.

22. P. W. Fowler, J. B. Sandall and S. J. Austin, *Fullerene Sci. & Technol.*, **4** (1996) 369; P. W. Fowler, J. B. Sandall and R. Taylor, *J. Chem. Soc., Perkin Trans. 2*, (1997) 419.

23. A. L. Balch, D. A. Costa, B. C. Noll and M. M. Olmstead, *J. Am. Chem. Soc.*, **117** (1995) 8926.

24. T. Hamano, T. Mashino and M. Hirobe, *J. Chem. Soc., Chem. Commun.*, (1995) 1537.

25. P. R. Birkett, P. B. Hitchcock, H. W. Kroto, R. Taylor and D. R. M. Walton, *Nature*, **357** (1992) 479.

26. F. N. Tebbe *et al.*, *Science*, **256** (1992) 822.

27. P. R. Birkett, A. G. Avent, A. D. Darwish, H. W. Kroto, R. Taylor and D. R. M. Walton, *J. Chem. Soc., Chem. Commun.*, (1993) 1260.

28. U. Reuther and A. Hirsch, *Chem. Commun.*, (1998) 1401.

29. P. J. Krusic, E. Wasserman, P. N. Keizer, J. R. Morton and K. F. Preston, *Science*, **254** (1991) 1183.

30. R. Taylor, *Electrophilic Aromatic Substitution*, Wiley, (1989) pp. 15–18.

31. P. R. Birkett, A. G. Avent, A. D. Darwish, H. W. Kroto, R. Taylor and D. R. M. Walton, *J. Chem. Soc., Perkin Trans. 2*, (1997) 457.

32. P. R. Birkett, A. G. Avent, A. D. Darwish, H. W. Kroto, R. Taylor and D. R. M. Walton, *J. Chem. Soc., Chem. Commun.*, (1995) 683.

33. P. J. Fagan, P. J. Krusic, C. N. McEwen, N. J. Lazar, D. H. Parker, N. Herron and E. Wasserman, *Science*, **262** (1963) 404; J. D. Crane, H. W. Kroto, G. J. Langley, R. Taylor and D. R. M. Walton, unpublished work; D. Brizzolara, J. T. Ahlemann, H. W. Roesky and K. Keller, *Bull. Soc. Chim. Fr.*, **130** (1993) 745.

34. A. L. Balch, V. J. Catalano, J. W. Lee, M. M. Olmstead and S. R. Parkin, *J. Am. Chem. Soc.*, **113** (1991) 8953.

35. A. Hirsch, T. Grösser, A. Skiebe and A. Soi, *Chem. Ber.*, **126** (1993) 1061.

36. X. Zhang, A. Romero and C. S. Foote, *J. Am. Chem. Soc.*, **115** (1993) 11024; X. Zhang, A. Fan and C. S. Foote, *J. Org. Chem.*, **61** (1996) 5456.

37. A. G. Avent *et al.*, *J. Chem. Soc., Perkin Trans. 2*, (1994) 15.

38. A. B. Smith *et al.*, *J. Chem. Soc., Chem. Commun.*, (1994) 2187.

39. M. F. Meidine *et al.*, *J. Chem. Soc., Perkin Trans. 2*, (1994) 1189.

40. M. S. Meier, M. Poplawska, A. L. Compton, J. P. Shaw, J. P. Selegue and T. F. Guarr, *J. Am. Chem. Soc.*, **116** (1994) 7044.

41. J. M. Hawkins, A. Meyer and A. Soi, *J. Am. Chem. Soc.*, **115** (1993) 7499.

42. C. C. Henderson, C. M. Rohlfing, K. T. Gillen and P. A. Cahill, *Science*, **264** (1994) 397.

43. S. R. Wilson and Q. Lu, *J. Org. Chem.*, **60** (1995) 6496.

44. A. D. Darwish, A. G. Avent, R. Taylor and D. R. Walton, *J. Chem. Soc., Perkin Trans. 2*, (1996) 2079.

45. A. Hermann, F. Diederich, C. Thilgen, H. ter Meer and W. H. Müller, *Helv. Chim. Acta*, **77** (1994) 1689.

46. O. V. Boltalina and R. Taylor, unpublished work.

4

Hydrogenation

Hydrogenation is the simplest reaction of fullerenes and was the first to be carried out; the hydrofullerenes can be converted back to [60]fullerene by heating with 2,3-dichloro-5,6-dicyanobenzoquinone (DDQ) in toluene.[1] Initially it was thought that it might be possible to add 60 hydrogen atoms to the [60]fullerene cage, but as noted in Chap. 3, this cannot be achieved because eclipsing interactions become severe, especially in a molecule that is particularly rigid. Reduction of the fullerenes takes place very readily, consequently they are good oxidising agents and for example [60]fullerene will oxidise hydrogen sulphide to sulphur.[2] The ease of hydrogenation makes the reaction difficult to control and it tends to produce $C_{60}H_{36}$ associated with, under some conditions, $C_{60}H_{18}$, these being particularly stable derivatives for reasons described in Sec. 4.2) A further problem is that the products tend to undergo very rapid allylic oxidation (especially when in solution), making structural analysis of them difficult.[3,4] Oxidation back to the parent fullerene can also occur,[5] and in general, oxidation is accelerated by the presence of unreacted [60]fullerene,[6] probably because the latter is a potent producer of singlet oxygen (and readily oxygenates alkylalkenes)[7] which a hydrofullerene effectively is. Because the fullerene hydrides are the simplest derivatives, theoreticians have been attracted by the *relative* simplicity of calculating the stabilities of the various isomers for a given addition level.

4.1 Hydrogenation Conditions

4.1.1 *Hydrogenation with Hydrogen Under Pressure*

Hydrogenation can be achieved most simply by heating the fullerene with hydrogen under pressure and at elevated temperatures. The level of hydrogenation can be controlled by adjusting the hydrogen pressure and the reaction temperature, and for example, a combination of 400°C and 80 atm. yield mainly $C_{60}H_{18}$;[8] up to 48 hydrogens can be added to [60]fullerene under more forcing conditions.[9] The additional advantage of this method is that there is no need to remove solvents on completion of reaction, and the absence of air at any stage coupled with the fact that the fullerene remains throughout as a solid (which reduces susceptibility towards oxidation) makes it the method of choice if the necessary equipment is available.

4.1.2 *Hydrogenation with Hydrogen and a Catalyst*

Various catalysts (Ru/carbon is particularly effective), have also been used to facilitate direct hydrogenation of [60]- and [70]fullerenes in toluene solution, giving e.g. $C_{60}H_{36}$, $C_{60}H_{18}$ and $C_{70}H_{38}$. Hydrofullerenes up to $C_{60}H_{50}$ were obtained using a temperature of 280°C and 160 atm. of hydrogen, though yields generally decreased with increasing temperature.[10] A peak at 780 amu, seen at higher temperatures, is not due to the formation of the $C_{60}H_{60}$ (*cf.* Ref. 10) but rather to the formation of methylene adducts arising from cage fragmentation, as discussed further below. Rhodium(0) on alumina catalyses hydrogenation of [60]fullerene in benzene to $C_{60}H_2$ in low yield,[5] as does Pt/C in hexane, hexene, or benzene.[3] Alkyl radicals can induce the formation of hydrogen atoms from hydrogen, and thus use of ethyl iodide/H_2 (6.9 Mpa) with [60]- and [70]fullerenes at 400°C produces a mixture of $C_{60}H_{36}$ and $C_{70}H_{36}$.[11]

4.1.3 *Dissolving Metal Reductions*

Next in order of simplicity are dissolving metal reductions. Hydrogenation of [60]fullerene under Birch conditions (Li/NH₃/t-BuOH) gives a mixture of $C_{60}H_{36}$ and $C_{60}H_{18}$,[1] but this method is complicated by the concurrent formation of amino derivatives due to the ease of nucleophilic addition to the cage.[12,13] More convenient and faster is reduction using Zn/HCl in toluene under N_2, and according to the acid strength, [60]fullerene gives either $C_{60}H_2$, $C_{60}H_4$, $C_{60}H_6$, or $C_{60}H_{36}$ and [70]fullerene mainly $C_{70}H_{38}$;[4,14] this is the best way to obtain the latter two derivatives free from impurities. Deuteriated fullerenes,[4] and hydrogenated [76]-, [78]- and [84]fullerenes[15] have been obtained in this way. Other metals that have been used are Sn, Al, Ti, Zn(Cu) and Mg.[16]

4.1.4 *Di-imide Reduction*

Hydrogenation using di-imide is a convenient method of producing dihydro- and tetrahydro derivatives of both [60]- and [70]fullerenes.[3] Hydrazine may also be used.[17]

4.1.5 *Reduction Using Organometallic Reagents*

Hydrogenation can be achieved by the use of organometallic reagents. Here initial formation of an intermediate H-fullerene-M (M = organometallic group) is followed by acid hydrolysis to give the dihydrofullerene. Examples are MH = $(\eta^5$-$C_5H_5)Zr(H)Cl$ (hydrozirconation)[13] and BH_3 (hydroboration),[18] the latter reaction being used for the first preparation and characterisation of $C_{60}H_2$ (10–30% yields) and for the preparation of $C_{70}H_2$ (20% yield).[19]

4.1.6 *Transfer Hydrogenation*

The principle behind the method is to react the fullerene with a compound that can readily lose hydrogen to become aromatic. Thus the transfer of hydrogen is energetically favourable. Transfer hydrogenation has the potential disadvantage of producing organic byproducts, removal of which may be both difficult and increase the likelihood of oxidation of the hydrofullerene product.

9,10-Dihydroanthracene is a typical reagent, and heating this with [60]fullerene in a sealed tube at 350°C for 30 min. gives $C_{60}H_{36}$, with $C_{60}H_{18}$ being formed after a longer reaction time (24 h). The formation of the latter is believed to be due to dehydrogenation by the anthracene byproduct (which forms 1,2,3,4-tetrahydroanthracene); the corresponding deuteriated fullerenes could be made by using deuteriated anthracene.[20] If 7*H*-benzanthrene is added to the mixture the reaction takes place at a lower temperature of 250°C but gives a wide range of products as does also the hydrogenation of [70]fullerene (which shows a maximum around 876 amu in the mass spectrum corresponding to $C_{70}H_{36}$).[21]

Just as cyclohexadiene readily transfers hydrogen and becomes oxidised to benzene, so dihydropyridines behave similarly and are converted to pyridine. (This is the principle behind the mode of action of NADH, the biological reducing agent). This technique (using ethyl 2,6-dimethyl-1,4-dihydropyridine-3,5-dicarboxylate) has been used to reduce [60]fullerene (in toluene at 100°C) to a mixture of $C_{60}H_n$, $n = 2,4,6,8,10$).[22]

1,2-Dihydro[60]fullerene can be prepared by irradiating [60]fullerene with 10-methyl-9,10-dihydroacridine in the presence of trifluoroacetic acid. Here the initially formed triplet excited state of [60]fullerene, is in turn converted to $C_{60}^{\bullet-}$, $C_{60}H^{\bullet}$, $C_{60}H^-$ and $C_{60}H_2$.[23]

Ultrasonic irradiation of [60]fullerene in the presence of decahydronaphthalene produces some $C_{60}H_2$, but the yield decreases with increased reaction times, and the parent fullerene also disappears for reasons not yet understood.[24]

4.2 Products of Hydrogenation

Some general points are considered first:

(i) The NMR resonances for hydrofullerenes are very solvent sensitive due to the acidity of the hydrogens, and thus appear *ca.* 1 ppm more downfield in CS_2 than in solvents such as benzene or toluene.[3]

(ii) The very downfield positions of the resonances indicate the strong electron-withdrawing effect of the cages.[3]

(iii) The positions are more downfield for hydro[60]fullerenes than are those for hydro[70]fullerenes, showing that [60]fullerene is the more strongly electron withdrawing.[3] This conflicts however with the greater electron affinity of [70]fullerene (gas-phase),[25] and the difference may therefore reflect solvation effects.

(iv) The cage should become less electron withdrawing as hydrogen is added to it (because sp³ hybridised carbons have a lower -I effect than sp²-hybridised carbons),[26] and results in an upfield shift for the resonances for the more hydrogenated fullerenes. Experimental results[3] and theoretical calculations[27] support this conjecture.

(v) The products of hydrogenation may be sensitive to the steric demands of the reducing reagent.

4.2.1 *Dihydrofullerenes*

(a) $C_{60}H_2$. Reaction of [60]fullerene under the conditions given in Sec. 4.1.5 produces 1,2-dihydro[60]fullerene, which is also the main product of reduction by di-imide (Sec. 4.1.4). The ^1H NMR spectrum shows a singlet at 5.93 ppm (toluene), and the magnitudes of D-H couplings in $C_{60}HD$ confirm the structure[18] The UV spectrum shows a peak around 434 nm (as does that produced by hydrozirconation), this being a characteristic of [60]fullerene 1,2-addition products.[28] 1,2-Dihydro[60]fullerene is also isolated from Zn/HCl reduction using 6 M HCl/toluene at 110°C.[14] and is found in the product of di-imide reduction.[3]

A second singlet observed in the ^1H NMR spectrum at 7.23 (CS_2), 6.31 (C_6H_6), comprising *ca.* 40% of the amount of the 1,2-isomer in di-imide reduction, and the major product in chromium acetate or Pt/H_2 reduction,[4] was provisionally attributed to 1,4-dihydro[60]fullerene,[3] but is 1,2,55,60 tetrahydro[60]fullerene. (below).

$C_{60}H_2$ undergoes three reversible reduction processes, with potentials *ca.* 0.1 V more negative that those for the corresponding electron additions to C_{60}. The strong electron-withdrawal by the cage makes $C_{60}H_2$ a strong acid (as strong as chloroacetic acid) and the first and second acid dissociation constants for $C_{60}H_2$ are 4.7 and 16, respectively.[29] The former value is consistent with that (5.7) for 1,2-*t*-butyl$C_{60}H$[30] since *t*-butyl is more electron supplying than H. The pK_a value for $C_{60}H^\bullet$ ($\rightarrow C_{60}H^{\bullet-} + H^+$) is *ca.* 9.[29]

(b) $C_{70}H_2$. [70]Fullerene gives two dihydro products, 1,2-dihydro[70]fullerene and 5,6-dihydro[70]fullerene[3] (formerly 1,9- and 7,8-dihydro[70]fullerenes), in relative yields varying from 8:1 (di-imide)[3] to 2:1 (hydroboration).[19] The additions are respectively along and across one of the symmetry planes, giving rise to a singlet at 4.96 (3.87) ppm in CS_2 (C_6D_6), and an AB quartet centred at 5.08 (3.96) ppm in CS_2 (C_6D_6).[3,19] As noted above, the resonances are upfield relative to those for $C_{60}H_2$, and the differential effects of the solvents may indicate differences in acidities of the 1,2- vs. the 5,6-hydrogens. The more downfield resonances for the hydrogens of the 1,2-dihydro isomer relative to those for the 5,6-isomer are attributed to the hydrogens being attached to the more electron-deficient pentagonal cap.[3]

Hydrogenation shows the pattern typical of reactions of [70]fullerene, namely that the 1,2-bond is more reactive than the 5,6-bond.

4.2.2 *Tetrahydrofullerenes*

Tetrahydro[60]fullerenes are produced in reduction by di-imide[3] and hydroboration[31] and the structures have been examined in detail; $C_{60}H_4$ has been produced also in hydrozirconation,[13] Zn/dil HCl,[14] Zn(Cu)aq.[16] and hydrazine[17] reductions. Eight different isomers are feasible (Fig. 4.1), and all are observed.[3] The large solvent shifts complicates comparisons between the two main studies[3,31] but the dominant product (and also in the hydrazine reduction) is 1,2,3,4-tetrahydro[60]fullerene (Fig. 4.1(a)), recognisable from the AA′BB′ ^{1}H NMR pattern, and this provides important information concerning the aromaticity of the cage. If the cage were completely non-aromatic, then this isomer would not be preferred, particularly in view of the eclipsing steric hindrance. However, the partial aromaticity is lost on reduction causing $x\pi$-electron localisation of the adjacent bonds so that further addition at these is preferred, (see also Sec. 3.1).

Of the eight isomers, the formation (from 1,2-$C_{60}H_2$) of seven of them (1,2,3,4; 1,2,7,21; 1,2,16,17; 1,2,18,36; 1,2,34,35; 1,2,33,50; 1,2,51,52) is favoured statistically (4:1) over the 1,2,55,60-isomer. The 1,2,3,4-isomer gives an AA′BB′ ^{1}H NMR pattern [6.02 (4.96) ppm in CS_2 (C_6D_6)], the 1,2,55,60 isomer gives a singlet [7.23 (6.31) ppm in CS_2 C_6D_6)], whilst the 1,2,18,36 isomer gives a singlet [6.27 (5.31) ppm, CS_2 (C_6D_6)] and an AB quartet [5.96 (5.03) ppm in CS_2 (C_6D_6)]; all the rest give AB quartets.[3,31] All are observed (together with an additional AB quartet due possibly to a higher hydrogenated derivative), six being present in greater quantity than the other two[3] as predicted by calculations;[31] the six most abundant isomers are also detected in [3]He NMR of a $C_{60}H_4$.[17] (Note that in this latter work, the chemical shifts are all 0.15 ppm downfield from those obtained using C_6D_6 as solvent).[3]

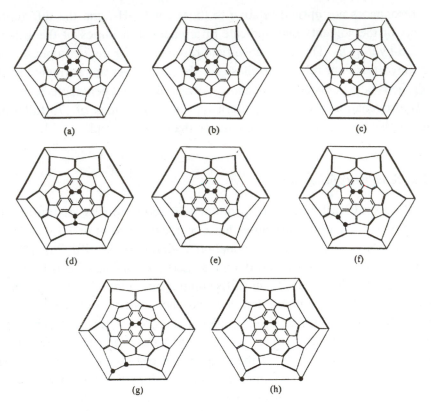

Fig. 4.1 Structures of the eight isomers of $C_{60}H_4$ (• = H).

An additional singlet observed in the di-imide reduction product at 6.38 (5.49) ppm in CS_2 (C_6D_6) and provisionally attributed to the 1,2,55,60 isomer,[3] is now unexplained.

Reduction by the zinc/copper couple[16] follows a different course, the main product being the 1,2,18,36 isomer (*e*) with the 1,2,33,50 isomer (*trans*-3) being a probable minor isomer. These results indicate significant steric effects with this reagent (see also Sec. 4.2.3).

Six tetrahydro[70]fullerenes are produced by di-imide reduction,[3] with the 1,2,3,4- and 1,2,5,6-isomers, Figs. 4.2(a), (b) (note numbering change from Ref. 3) each comprising *ca.* 25% of the total, and giving AA'BB' and ABCD ^{1}H NMR patterns, respectively. The preference for these derivatives follows the same argument that accounts for the preferential formation of

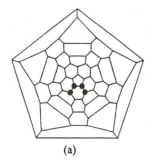
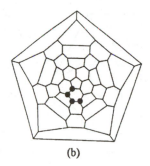

(a)   (b)

Fig. 4.2 Structures of 1,2,3,4-tetrahydro[70]fullerene and 1,2,5,6-tetrahydro[70]fullerene (• = H).

1,2,3,4-tetrahydro[60]fullerene. Although initial addition across the 1,2-bond is favoured over addition across the 5,6-bond (Sec. 4.2.1), subsequent 1,2-addition to the 5,6-dihydro compound has a 2:1 statistical advantage over subsequent 5,6-addition to the 1,2-dihydro compound, so that similar amounts of each that are obtained.[3] In both compounds the more downfield resonance is attributed to the hydrogen(s) attached to the pentagonal cap, and in the 1,2,5,6 compound the resonance for this hydrogen is substantially (0.8 ppm) downfield from the rest.

The remaining four isomers are likely to be each of the possibilities (1,2,13,30; 1,2,41,58; 1,2,56,57; 1,2,59,60) that follow from further addition to 1,2-dihydro[70]fullerene, since yields of derivatives derived addition in the other polar region to the low yield 5,6-dihydro precursor should be very small.[3]

4.2.3 $C_{60}H_6$

$C_{60}H_6$ has been detected in four studies,[13,14,16,32] but the characterisation as the 1,2,33,41,42,50 derivative (**4.1**)[16,32] in reduction by Zn(Cu)aq. presents a problem of rationalization, since (a) it cannot be obtained by further hydrogenation of the main 1,2,3,4-tetrahydro precursor, (b) it is not on the pathway to the formation of $C_{60}H_{18}$, the next highest (and fully characterised) hydrogenation level, and (c) calculations[33] do not predict it to be a stable derivative. Instead these predict that both the 'S'-shaped addition pattern

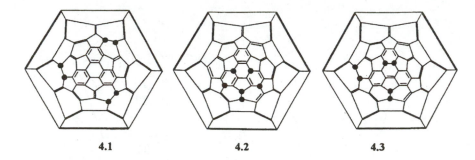

4.1	**4.2**	**4.3**

(Fig. 3.1) and that found in 6-fold chlorination and bromination (**4.2**), should occur with equal probability. The results indicate that reduction by the Zn(Cu) couple has different steric requirements from other hydrogenation methods. A minor isomer is believed to be (**4.3**), this addition pattern being predicted in early work based on the retention of a decacyclene sub-structure,[34] as is the case here.

4.2.4 $C_{60}H_{18}$

This is a red solid that can be made by direct hydrogenation of [60]fullerene at high temperature and pressure.[8] It has been fully characterised[35] and has a C_{3v} symmetry crown structure (Fig. 3.8, • = H) predicted by calculations to be particularly stable.[36] It is isostructural with $C_{60}F_{18}$ (Sec. 7.1.1.2) and on standing in air over an extended period, oxide derivatives (as yet uncharacterized) are formed. In solution, degradation is rapid.[35]

4.2.5 $C_{60}H_{36}$

This is obtained in many reductions (Sec. 4.1.1, 4.1.2, 4.1.3, 4.1.6), but is most easily produced in the laboratory by stirring a solution of the fullerene in toluene with Zn/conc. HCl. Reduction is very rapid and complete within 1 h.

Although the ready formation of $C_{60}H_{36}$ (often accompanied by the formation of $C_{60}H_{18}$) indicates a particularly stable structure, no analysable ^{1}H NMR spectrum has been obtained, due to the very rapid oxidation of the material when in solution.[3] The first conjectured structure[1] was improbable since it had all-isolated double bonds. An alternative tetrahedral (T symmetry) structure (Fig. 3.9) proposed by the writer[37] was indicated by theoretical calculations to be very stable.[38,39] More recently the importance of C_3 (Fig. 3.10), D_{3d} and S_6 structures has also been emphasized.[40–43] [Note that there are many isomers having the same symmetry.[41] Thus the D_{3d} isomer (isomer No. 1 of the comprehensive list given in Ref. 41) is described as D_{3d}' in Ref. 42; the isomers described as D_{3d} and as S_6 in Refs. 40, 42 and 44 are of higher energy since they have six double bonds in pentagons, a destabilising feature, and are not listed in Ref. 41.]

AM1/MOPAC calculations indicate the order of isomer stabilities as $D_{3d} \approx T >$ two C_3 isomers, the difference between the two sets being *ca.* 30 kcal/mole.[41] All but one of the C_3 isomers contain the $C_{60}H_{18}$ motif as a subset (compare Figs. 3.9 and 3.10 with 3.8) which is relevant to the fact that heating $C_{60}H_{36}$ to 650°C causes slow degradation to $C_{60}H_{18}$.[4] All of these isomers contain one or more benzenoid rings in the structure, and notably, the UV spectrum of $C_{60}H_{36}$ is consistent with the presence of isolated benzenoid structures.[44] Indirect information on the structure comes from fluorination (Chap. 7), which parallels hydrogenation: $C_{60}F_{36}$ consists of two isomers having T and C_3 symmetry in *ca.* 1:3 ratio. The indication that $C_{60}H_{36}$ is similarly comprised comes from ^{3}He NMR of i-^{3}HeC$_{60}$H$_{36}$, which shows that only two isomers are present, and in *ca.* 2:1 ratio.[17] Both of these structures can be built up from any of the eight tetrahydro[60]fullerenes.

Deuteration of [60]fullerene using Zn/DCl gives a higher level of reduction, with products up to $C_{60}D_{44}$ being obtained. This may be due to the higher stability of the C-D bond compared to the C-H bond, so providing better compensation for both the loss of aromaticity, and the increase in steric compression that accompanies reduction beyond the 36 H level.[4]

4.2.6 $C_{70}H_{8/10/12}$

Hydrogenation of [70]fullerene with Zn(Cu)aq, produces three isomers of $C_{70}H_{10}$ together with smaller amounts of $C_{70}H_8$ and $C_{70}H_{12}$,[16] but these derivatives have not been characterised so far; $C_{70}H_{12}$ was first detected as a spurious product in HPLC separation of higher fullerenes using hexane as eluent.[45]

4.2.7 $C_{70}H_{36/38/40}$

Hydrogenation of [70]fullerene by Zn/HCl is slower than for [60]fullerene and gives a mixture of $C_{60}H_{36/38/40}$ with $C_{70}H_{38}$ the main component; fluorination here too gives the same addition pattern (Chap. 7).[4] The structures of these isomers have not been obtained, but proposals based on creating a large equatorial region of enhanced aromaticity have been made,[21,39,46] though calculations indicate that the cage curvature may limit this.[47] The UV spectrum of the reduced [70]fullerene showed bands indicative of some aromatic character. These hydrogenated derivatives are not so thermally stable as $C_{60}H_{36}$ and degradation goes directly to [70]fullerene, with no intermediate product corresponding to $C_{60}H_{18}$ being formed.[4]

The behaviour of [60]fullerene towards Zn/DCl (above) is paralleled, with a maximum reduction level of $C_{70}D_{48}$ being obtained.[4]

4.2.8 *Hydrogenated Higher Fullerenes*

Hydrogenation of a mixture of [76]-, [78]- and [84]fullerenes with Zn/HCl gives, respectively, $C_{76}H_{46-50}$, $C_{78}H_{36-48}$ and $C_{84}H_{48-52}$. The broader spectrum of products from [78]fullerene may reflect the diverse isomeric composition of the parent fullerene. Two notable features of these results are: (i) hydrogenated [60]- and [70]fullerene are obtained as byproducts in each case, indicating substantial cage breakdown, and moreover extended (5 h) reduction of [84]fullerene showed *complete* breakdown to $C_{60}H_{36}$;[15] (ii) at high mass spectrometry probe temperatures, the product from reaction of [84]fullerene is mainly $C_{84}H_{40}$ which again parallels the addition level in fluorination.[48]

4.2.9 *Formation of Methylene Adducts*

Hydrogenation is accompanied by the formation of methylene adducts, in particular trimethylene adducts showing 42 amu mass increments. Observed species of 780, 798 and 840 amu are assignable to addition of one or more trimethylene moieties to $C_{60}H_{18}$ and $C_{60}H_{36}$.[4] In the light of the fragmentation of the higher fullerenes noted above, it is most probable that these species are formed from reductive cleavage of the fullerene cages, but no further information is presently available.

4.3 Theoretical Calculations

(a) $C_{60}H_2$ and $C_{70}H_2$. Calculations of the energies of all of the 23 possible dihydro[60]fullerenes, show that only three are significantly stable, these being the 1,2- > 1,4- > and 1,16-isomers.[49] The results originate from the unfavourability of placing double bonds in pentagons (these three isomers require 0, 1, and 2 so placed, respectively), and the further the hydrogens are placed apart, the more this feature is necessitated; more extensive calculations confirm this.[50] There is a remarkable parallel between these results and the order of conjugative interactions in naphthalene which follow the same order *viz.* 1,2- > 1,4- > 2,6 (equivalent to the 1,16 interaction). In naphthalene this order is dictated similarly by the need to conserve the bond fixation in the ground state.[51]

Similar calculations for dihydro[70]fullerenes predict the order of addition to be 5,6 > 1,2 > 7,23.[51] The latter prediction is significant in the context of chlorination (Chap. 7). The first two orders (also predicted by other calculations)[52] are the reverse of observations in hydrogenation as well as other reactions and is thought to be due to differences in curvature of the cage surface across the two sites.

(b) $C_{60}H_4$. At these higher levels of hydrogenation, kinetic factors tend to render the thermodynamically based theoretical calculations less reliable. Thus the 1,2,3,4-tetrahydro[60]fullerene is the most readily formed because of bond-localisation factors (Sec. 4.2.2) but is predicted to be almost the least stable[51] presumably because of destabilising eclipsing factors, nevertheless the values for all of the isomers are very close.

(c) $C_{60}H_6$. Calculations for hexahydro[60]fullerenes derived from a 1,2-dihydro precursor indicate that the three most stable derivatives (in order) are: 1,2,4,11,15,30 > 11,2,3,4,9,10 > 1,2,3,4,11,12.[33] The latter two are the 'S' and 'T' derivatives, respectively (see Fig. 3.1), expected from kinetic and eclipsing considerations. The predicted higher stability of the former isomer (**4.3**) is however surprising, and although there is no evidence as yet for this addition pattern in hydrogenation, it is observed in both bromination and chlorination (Chap. 7), and also in the addition of morpholine.[53] The fundamental reason for the stability is presently unclear.

(d) $C_{60}H_{36}$. Calculations of the possible structures for this compound are described above (Sec. 4.2.5).

References

1. R. E. Haufler *et al.*, *J. Phys. Chem.*, **94** (1990) 8634.
2. A. D. Darwish, H. W. Kroto, R. Taylor and D. R. Walton, *Fullerene Sci. & Technol.*, **1** (1993) 571.
3. A. G. Avent *et al.*, *J. Chem. Soc.*, *Perkin Trans. 2*, (1994) 15.
4. A. D. Darwish, A. K. Abdul-Sada, G. J. Langley, H. W. Kroto, R. Taylor and D. Walton, *J. Chem. Soc.*, *Perkin Trans. 2*, (1995) 2359.
5. L. Becker, T. P. Moore and J. L. Bada, *J. Org. Chem.*, **58** (1993) 7630.
6. C. C. Henderson and P. A. Cahill, *Science*, **259** (1993) 1885.
7. M. Orfanopoulos and S. Kambourakis, *Tetrahedron Lett.*, **35** (1994) 1945.
8. Y. Sui *et al.*, *Fullerene Sci. & Technol.*, **4** (1996) 813.
9. R. Loutfy, personal communication.
10. K. Shigematsu and K. Abe, *Chem. Express*, **7** (1992) 905; K. Shigematsu, K. Abe, M. Mitani and K. Tanaka, *ibid.*, **8** (1993) 37, 483.
11. M. I. Attalla, A. M. Vassallo, B. N. Tattam and J. V. Hanna, *J. Phys. Chem.*, **97** (1993) 629.
12. R. Taylor, *Philos Trans. Roy. Soc. (London)*, **A343** (1993) 87.
13. S. Ballenweg, R. Gleiter and W Krätschmer, *Tetrahedron Lett.*, (1993) 3737.
14. M. S. Meier, P. S. Corbin, V. K. Vance, M. Clayton, M. Mollman and M. Poplawska, *Terahedron Lett.*, **32** (1994) 5789.
15. A. D. Darwish, H. W. Kroto, R. Taylor and D. R. M. Walton, *J. Chem. Soc.*, *Perkin Trans. 2*, (1996) 415.
16. R. G. Bergosh, M. S. Meier, J. A. Laske-Cooke, H. P. Spielmann and B. R. Weedon, *J. Org. Chem.*, **62** (1997) 7667.
17. W. E. Billups *et al.*, *Tetrahedron Lett.*, **38** (1997) 171, 175.
18. C. C. Henderson and P. A. Cahill, *Science*, **259** (1993) 1885.

19. C. C. Henderson, C. M. Rohlfing, K. T. Gillen and P. A. Cahill, *Science*, **264** (1994) 397.

20. C. Rüchardt *et al.*, *Angew. Chem. Intl. Edn. Engl.*, **32** (1993) 584.

21. M.Gerst, H.-D. Beckhaus, C. Rüchardt, E. E. B. Campbell and R. Tellgmann, *Tetrahedron Lett.*, **34** (1993) 7729.

22. N. F. Gol'dschleger *et al.*, *Russ. Chem. Bull.*, **45** (1996) 2402.

23. S. Fukuzumi, T. Suenobu, S. Kawamura, A. Ishida and K. Mikami, *Chem. Commun.*, (1997) 291.

24. D. Mandrus, M. Kele, R. L. Hettich, G. Guiochon, B. C. Sales and L. A. Boatner, *J. Phys. Chem.*, **B101** (1997) 123.

25. O. V. Boltalina, E. V. Dashkova and L. N. Sidorov, *Chem. Phys. Lett.*, **256** (1996) 253.

26. R. Taylor and D. R. Walton, *Nature*, **363** (1993) 685.

27. K. Chono, G. Van Lier, G. Van de Woude and P. Geerlings, *J. Chem. Soc.*, *Perkin Trans. 2*, (1998) 1723.

28. H. R. Karfunkel and A. Hirsch, *Angew. Chem. Intl. Edn. Engl.*, **31** (1992) 1468.

29. M. E. Niyazymbetov, D. H. Evans, S. A. Lerke, P. A. Cahill and C. C. Henderson, *J. Phys. Chem.*, **98** (1994) 13093.

30. P. J. Fagan, P. J. Krusic, D. H. Evans, S. A. Lerke and E. Johnstone, *J. Am. Chem. Soc.*, **114** (1992) 9697.

31. C. C. Henderson, C. M. Rohlfing, R. A. Assink and P. A. Cahill, *Angew. Chem. Intl. Edn. Engl.*, **33** (1994) 786.

32. M. S. Meier, B. R. Weedon and H. P. Spielmann, *J. Am. Chem. Soc.*, **118** (1996) 11682.

33. P. A. Cahill, *Chem. Phys. Lett.*, **254** (1996) 257; P. A. Cahill and C. M. Rohlfing, *Tetrahedron*, **52** (1996) 5247.

34. R. Taylor, *J. Chem. Soc.*, *Perkin Trans. 2*, (1992) 1667.

35. A. D. Darwish, A. G. Avent, R. Taylor and D. R. M. Walton, *J. Chem. Soc.*, *Perkin Trans. 2*, (1996) 2079.

36. B. W. Clare and D. L. Kepert, *J. Mol. Struct.* (*THEOCHEM*), **303** (1994) 1.

37. R. Taylor, *J. Chem. Soc.*, *Perkin Trans. 2*, (1992) 1667; *Philos. Trans. Roy. Soc. London Ser.*, **A343** (1993) 87.

38. S. J. Austin, R. C. Batten, P. W. Fowler, D. B. Redmond and R. Taylor, *J. Chem. Soc.*, *Perkin Trans. 2*, (1993) 1383; B. I. Dunlap, D. W. Brenner,

J. W. Mintmire, R. C. Mowery and C. T. White, *J. Phys. Chem.*, **95** (1991) 5763; A. Rathna and J. Chandrasekhar, *Chem. Phys. Lett.*, **206** (1993) 217.

39. L. D. Book and G. E. Scuseria, *J. Phys. Chem.*, **98** (1994) 4283.

40. B. I. Dunlap, D. W. Brenner and G. W. Schriver, *J. Phys. Chem.*, **98** (1994) 1756.

41. B. W. Clare and D. L Kepert, *J. Mol. Struct. (THEOCHEM)*, **315** (1994) 71.

42. M. Bühl, W. Thiel and U. Schneider, *J. Am. Chem. Soc.*, **117** (1995) 4623.

43. L. E. Hall *et al.*, *J. Phys. Chem.*, **97** (1993) 5741.

44. R. V. Bensasson *et al.*, *Chem Phys.*, **215** (1997) 111.

45. R. Taylor, G. J. Langley, A. G. Avent, T. J. S Dennis, H. W. Kroto and D. M. Walton, *J. Chem. Soc., Perkin Trans. 2*, (1993) 1029.

46. R. Taylor, *J. Chem. Soc., Perkin Trans. 2*, (1994) 2497.

47. P. W. Fowler, J. P. B. Sandall, S. J. Austin, D. E. Manolopoulos, P. D. M. Lawrenson and J. M. Smallwood, *Synthetic Metals*, **77** (1996), 97; P. W. Fowler, J. P. B. Sandall and S. J. Austin, *Fullerene Sci. Tech.*, **4** (1996) 369; P. W. Fowler, J. P. B. Sandall and R. Taylor, *J. Chem. Soc., Perkin Trans. 2*, (1997) 419.

48. O. V. Boltania and R. Taylor, unpublished work.

49. C. C. Henderson and P. A. Cahill, *Chem. Phys. Lett.*, **198** (1992) 570; C. C. Henderson, C. M. Rohlfing and P. A. Cahill, *ibid.*, **213** (1993) 383.

50. B. W. Clare and D. L. Kepert, *J. Mol. Struct. (THEOCHEM)*, **281** (1993) 45.

51. R. Taylor, *Electrophilic Aromatic Substitution*, Wiley, Chichester, 1989, pp. 105–110.

52. V. N. Bezmelnitsin, A. V. Eletskii, N. G. Schepetov, A. G. Avent and R. Taylor, *J. Chem. Soc., Perkin Trans. 2*, (1997) 683.

53. G. Schick, K.-D. Kampe and A. Hirsch, *J. Chem. Soc., Chem. Commun.*, (1995) 2023.

5

Reduction by Electron Addition, and Reaction of Fullerene Radical Anions with Electrophiles

The term *reduction* includes any reaction involving the addition of electrons, and such processes involving fullerenes are described here. Fullerenes are unique in respect of the large number of electrons that can be added to them. Addition of electron pairs accompanied by covalent bond formation to another group also occurs and is described under nucleophilic addition in Chap. 6. Electrophiles do not react at all readily with fullerenes because of the electron deficiency of the cage. This obstacle to the preparation of many important derivatives can be overcome by first preparing either radical anions, or anions, which then react readily with electrophiles. So far, these important routes have been exploited but little; reaction of electrophiles with radical anions is described in this chapter.

5.1 Formation of Radical Anions

Addition of a single electron (or indeed an odd number of electrons) to a fullerene produces a *radical anion*, e.g. $C_{60}^{\bullet-}$, described as e.g. a fullerene-1-elide when the radical is located at the 1-position. When an even number of electrons (e.g. two) are added, the even number of free electrons may combine to give bond, so that a pure *anion*, (in this case a *dianion*) is obtained; if combination does not occur, then a di-*radical anion* is obtained. There is a high probability of the latter because delocalisation of the electrons across the cage is unfavourable since it produces double bonds in pentagons;

71

delocalisation is therefore local rather than global. This underlies the general observation that the regiochemistry in one area of the cages is largely unaffected by addends located elsewhere. (NB. In many publications, it is assumed that the free electrons have combined, and therefore $C_{60}^{2\bullet-}$ is written as C_{60}^{2-}.)

Electrons can be added reversibly to fullerenes by reduction, either electrochemically, or, for example, with sodium in liquid ammonia. The LUMO of [60]fullerene is low lying and triply degenerate, and is therefore readily reduced to the hexa-anion; the presence of very many sp^2-hybridised carbons also contributes to the electronegativity of the molecule. The acquisition of six electrons is related also to the presence of six pyracyclene units in [60]fullerene, and addition of one electron to each (Fig. 5.1) produces a five-membered 6π aromatic ring. Six pyracyclene units are also present in [70]-, [76]- [78]- and [84]fullerenes, and so addition of up to six electrons to each can be achieved,[1,2] though there is evidence from alkylation (Sec. 5.6) that very many more electrons (> 30) can be accommodated on the cages, a remarkable phenomenon. As expected, the radical anion formed on addition of one electron to [60]fullerene gives rise to an ESR signal,[3] with a g value of e.g. 2.009 for that formed from [60]fullerene and Na/THF at low temperature[4] (the line widths increase with increasing temperature); the monoanions typically absorb in the near infra red at 1068 nm.[5] Each radical anion has a distinctive colour, e.g. $C_{60}^{\bullet-}$, $C_{60}^{2\bullet-}$ and $C_{60}^{3\bullet-}$ are dark red-purple, orange-red and dark red-brown, respectively.[6]

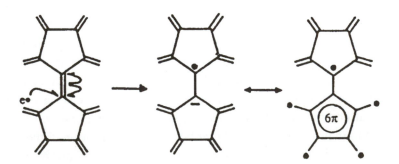

Fig. 5.1 Addition of an electron to a pyracyclene unit.

A consequence of the ready formation of radical anions is that they are stable, and solutions of them may be kept for many days. They are soluble in, e.g. THF, and suspensions of the fullerenes in this solvent clarify as reduction proceeds. Formation of the radical anions is more difficult if electron-supplying groups are bonded to the cages and hence the potential of the first reduction wave (measured by cyclic voltammetry) is increased (becomes more negative).[7] The reduction potentials are depend somewhat upon the electron-donor/acceptor properties of the solvent,[6] being more negative the higher the acceptor property though the reasons for this are unclear. [70]Fullerene accepts electrons more readily than [60]fullerene [which is consistent with its higher electron affinity (Sec. 2.7)] and the hexa-anion is more stable.[1d,e,2] This has been quantified in that formation of the monoanions from reaction of fullerenes with sodium in toluene occurs 2.5 times faster for [70]fullerene than for [60]fullerene.[8]

Many methods for making radical anions have been investigated and have been directed towards controlling the number of electrons added:

5.2 Reduction by Metals

The use of sodium has already been mentioned. Reduction can be carried out with lithium in THF (aided with ultrasound),[9] potassium in liquid ammonia (here the potassium salts of C_{60}^{n-}, $n = 1-4$, dissolve in the solvent, whereas those of $n = 5,6$ are insoluble)[2] and by mercury in a THF solution containing tetrahexylammonium bromide (creating a $Hg/HgBr_2$ couple) which produces the monoanion with $< 5\%$ of the dianion present.[10] Reduction by this less electropositive metal exemplifies the ease with which fullerenes can be reduced. Likewise reduction by Zn/NaOH in THF (or DMSO) may be used to prepare both mono- and dianions.[11]

Monoanionic salts $[C_{60}]^{\bullet-}K^+$ (g 1.9987) can be obtained in 94% yield very conveniently by reduction with K/THF in the presence of 1-methylnaphthalene,[12] and $Na^+[C_{60}]^{\bullet-}(THF)_5$ (g 1.999) is produced by reducing [60]fullerene with $Na[Mn(\eta\text{-}C_5Me_5)_2]$.[13] $C_{60}^{n\bullet-}$ species ($n = 2$ or 3) are obtainable by reaction of [60]fullerene with Na/THF in the presence of either two or three equivalents of dibenzo[18]crown-6, respectively,[14] whilst $C_{60}^{4\bullet-}$ results from reduction of [60]fullerene with Na/cryptand 222 in THF.[15] Selective formation of either $C_{60}^{\bullet-}$ or $C_{60}^{2\bullet-}$ can be obtained by reducing

[60]fullerene with Al-Ni alloy in either NaOH-THF or NaOH-DMSO (or DMF), respectively.[16]

5.3 Reduction by Organic Donors

Here an organic or organometallic molecule (instead of a metal) supplies the electrons to the cage, and gives rise to charge-transfer complexes. These are generally soluble in polar solvents e.g. THF or PhCN, and insoluble in non-polar ones e.g. hexane, so that addition of the latter to a solution in the former results in precipitation. Reagents that have been used to produce $C_{60}^{\bullet-}$ include THF/toluene solutions of $Cr^{II}PPh_3$ (which oxidises to $Cr^{III}PPh_3$, g for the anion being 1.995),[17] cobalticene and N-methylimidazole/tetraporphinatotin^{IV}.[18] By choice of the appropriate stoicheiometry, either one, two or three electrons can be transferred to [60]fullerene from the complex $[Fe^I(C_5H_5)(C_6Me_6)](X)$, the g values being 2.002 for the singly and double-charged species (linewidths 2.6 G and 4.0 G respectively at 300 K), and 2.004 for the triply charged one (linewidth 46 G);[19] in general, g values for the fullerene radical ions increase with increasing ion charge. $C_{60}^{n\bullet-}$, $n = 1,2$, have been obtained by using p-quinone dianions as the reductant.[20]

Interestingly, the diradical anion $C_{60}^{2\bullet-}$ formed by reaction with $(Ph_3P)_2NCl$ in MeCN, is slightly elongated, due probably to mutual repulsion of the charges located on opposite sides of the molecule.[21] The complex $[TDAE^+]C_{60}^{\bullet-}$ formed from [60]fullerene and tetrakis(dimethylamino)ethene is notable for undergoing transition to a ferromagnet at 16.1 K, and although this is *ca.* 30 times higher than for any previous organic molecular ferromagnet, it shows no remanance.[22]

5.4 Electrocrystallisation

The aim of this technique (which is an extension of the methods described in Sec. 5.3) is to produce radical anions with a specific charge, through their precipitation from solution. As indicated in Sec. 5.2, solubility decreases with increasing reduction level, and this is due to the increasing bulk of the anion-cation complex. Hence by using bulky cations, selective precipitation at a particular charge level may be achieved. Thus far, salts obtained in this way include $(Ph_4P^+)_3(C_{60}^{\bullet-})(Cl^-)_2$ (which behaves as a semiconductor at room temperature),[23] $[(Ph_3P)_2N]^+ \cdot (C_{60}^{\bullet-})$,[24] $(Ph_4P^+)_2(C_{60}^{\bullet-})Cl^-$ (here the fullerene lies at the centre of a P_8 cube),[25] and $(Ph_4P^+)_2(C_{60}^{\bullet-})(I^-)_n$ ($n = 0, 1$);[26] electrocrystallisation using tetraphenylarsonium halides has also been achieved.[27]

5.5 Electrochemical Reduction

The above methods all suffer from the disadvantage of the presence of additional reagents, which complicates subsequent work-up (already a formidable problem in fullerene chemistry). Cyclic voltammetry studies showed that electrons can be added stepwise and reversibly to fullerenes, the radical anions having stability windows of *ca.* 0.5 v.[1c–e,2] Thus by controlling the applied potential, the desired radical anion can then be produced, and reacted with various electrophiles as described below. This method may become a significant synthetic route for fullerene derivatisation.

5.6 Reaction of Electrophiles with Radical Anions

Controlled electrochemical reduction of [60]fullerene followed by reaction with methyl iodide produced $C_{60}Me_2$ consisting of a mixture of two isomers, presumed to be the 1,2- and 1,4-dimethyl derivatives (Fig. 5.2, R = Me),[28] and similar control leads to the formation of $C_{60}R_2$ (R = Et, n-Bu), $C_{60}Me_4$, and $C_{60}Me_6$.[29] By contrast, benzyl groups add to $C_{60}^{2\bullet-}$ only across the 1,4-positions (Fig. 5.2(b), R = CH_2Ph),[30] and this may be attributed to steric hindrance. Steric effects likewise account for addition of α,α'-dibromo-*o*-xylene occurring across the 1,2-positions only,[20] and permit the 1,2-addition of H and $PhCH_2$.[31] $C_{60}H_2$ has been prepared by reaction of protons from trifluoromethansulphonic acid with electrochemically produced $C_{60}^{2\bullet-}$,[32] but no reaction occurs with the weaker trifluoroacetic acid.[33]

Less controlled is the reaction of [60]fullerene with lithium to give the radical anion, followed by reaction with methyl iodide, whereby up to *ca.* 24 methyl groups become attached to the cage, with unknown locations.[34] Notably, $C_{60}Me_6$ and $C_{60}Me_8$ species were dominant in the spectrum suggesting that they may be isostructural with the bromo compounds (Chap. 7). The easier transfer of electrons from potassium results in the addition of up to 32 methyl

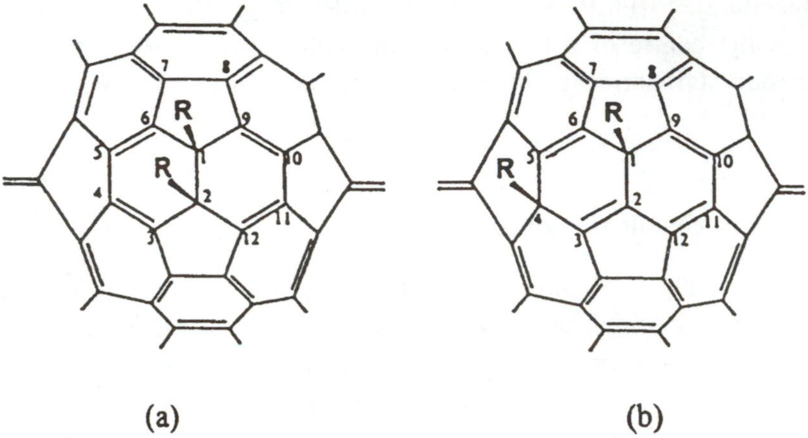

(a)   (b)

Fig. 5.2 Addition of alkyl groups at (a) the 1,2- and (b) the 1,4-sites in [60]fullerene.

groups to the cage (after reaction of the intermediate anionic species with MeI), and $C_{60}Me_6$ is also a prominent species in the sublimed product.[35] The reaction with the methyl iodide is instantaneous, and no further anions can be formed at this stage, hence 32 electrons must be transferred (at least partially) from the metal to the cage, a remarkable phenomenon.

5.7 Reaction of Fullerenes with Alkali- and Alkaline Earth Metals

Heating mixtures of solid [60]fullerene and the alkali metals (the reaction may also be carried out in either toluene[36] or liquid ammonia[37]) results in intercalation of the metal into the fullerene lattice, producing compounds of the general formula $C_{60}M_n$, where M is either Na, K, Rb, Cs (or combinations of these), and n can variously be 2,3,4 or 6. They may also be produced by oxidation of hydrides or borohydrides by [60]fullerene, hydrogen being liberated from the hydride.[38] These compounds have attracted enormous interest, because some of the compounds for $n = 3$ are superconductors.[39] However, they are pyrophoric, so exploitation of this property may be severely limited.

The A_3C_{60} compounds are face-centred cubic but become body-centred cubic at 40 K. K_3C_{60} has metallic/semiconductor properties and its 6:6- and

6:5-bond lengths are 1.444 and 1.434 respectively,[40] so it is more aromatic than the parent fullerene (*cf*. Sec. 2.7).

AC_{60} (A = K, Cs, Rb) compounds are air stable and metallic; K_4C_{60} and K_6C_{60} are insulators. Intercalation of either barium or calcium has also been achieved by heating the metal with [60]fullerene, giving Ca_5C_{60} and Ba_6C_{60} both of which also exhibit superconducting behaviour.[41]

References

1. (a) P.-M. Allemand *et al.*, *J. Am. Chem. Soc.*, **113** (1991) 1050; (b) D. M. Cox, *et al.*, *J. Am. Chem. Soc.*, **113**, (1991) 2940; (c) D. Dubois, K. M. Kadish, S. Flanagan and L. J. Wilson, *J. Am. Chem. Soc.*, **113** (1991) 7773; (d) Y. Ohsawa and T. Saji, *J. Chem. Soc., Chem. Commun.*, (1992) 781; (e) Q. Xie, E. Perez-cordero and L. Echegoyen, *J. Am. Chem. Soc.*, **114** (1992) 3978; (f) M. S. Meier, T. F. Guarr, J. P. Selegue and V. K. Vance, *J. Chem. Soc., Chem. Commun.*, (1993) 63; (g) Q. Li, F. Wudl, C. Thilgen, R. L. Whetten and F. Diederich, *J. Am. Chem. Soc.*, **114** (1992) 3996; (h) J. P. Selegue, J. P. Shaw, T. F. Guarr and M. S. Meier, *Recent Advances in the Chemistry and Physics of Fullerenes* (The Electrochem. Soc.) Eds. K. M. Kadish and R. S. Ruoff, **94–24** (1994) 1274.
2. F. Zhou, C. Jehoulet and A. J. Bard, *J. Am. Chem. Soc.*, **114** (1992) 11004.
3. P.-M. Allemand *et al.*, *J. Am. Chem. Soc.*, **113** (1991) 2780; M. A. Greaney and S. M. Gorun, *J. Phys. Chem.*, **95** (1991) 7142; D. Dubois, M. T. Jones and K. M. Kadish, *J. Am. Chem. Soc.*, **114** (1992) 6446.
4. S. P. Solodnikov, V. V. Bashilov and V. I Sokolov, *Bull. Acad. Sci. USSR*, **41** (1992) 2234.
5. D. R. Lawson *et al.*, *J. Electrochem. Soc.*, **139** (1992) L68.
6. L. J. Wilson, S. Flanagan, L. P. F. Chibante and J. M. Alford, *Buckminsterfullerenes*, Eds. W. E. Billups and M. A. Ciufolini, VCH, New York, (1993) 285.
7. S. A. Lerke, B. A. Parkinson, D. H. Evans and P. J. Fagan, *J. Am. Chem. Soc.*, **114** (1992) 7807.
8. V. N. Bezmelnitsyn, A. A. Dityatev, V. Ya. Davydov, N. G. Shepetov, A. V. Eletskii and V. F. Sinyanskii, *Chem. Phys. Lett.*, **237** (1995) 246.

9. J. W. Bausch, G. K. S. Prakash, D. S. Tse, D. C. Lorents, Y. K. Bae and R. Malhotra, *J. Am. Chem. Soc.*, **113** (1991) 3205.

10. P. Boulas, R. Subramanian, W. Kutner, M. T. Jones and K. M. Kadish, *J. Electrochem. Soc.*, **140** (1993) L130.

11. M. Wu, X. Wei, L. Qi and Z. Xu, *Tetrahedron Lett.*, (1996) 7409.

12. J. Chen, Z. Huang, R. Cai, Q. Shao, S. Chen and Y. He, *J. Chem. Soc., Chem. Commun.*, (1994) 2177.

13. R. E. Douthwaite, A. R. Brough and M. L. H. Green, *J. Chem. Soc., Chem. Commun.*, (1994) 267.

14. P. Bhyrappa, P. Paul, J. Stinchcombe, P. D. W. Boyd and C. A. Reed, *J. Am. Chem. Soc.*, **115** (1993) 11004.

15. Y. Sun and C. A. Reed, *Chem. Commun.*, (1997) 747.

16. X. Wei, M. Wu, L. Qi and Z. Xu, *J. Chem. Soc., Perkin Trans. 2*, (1997) 1389.

17. A. Pénicaud *et al.*, *J. Am. Chem. Soc.*, **113** (1991) 6698.

18. J. Stinchcombe, A. Pénicaud, P. Bhyrappa, P. D. W. Boyd and C. A. Reed, *J. Am. Chem. Soc.*, **115** (1993) 5212.

19. C. Bossard *et al.*, *J. Chem. Soc., Chem. Commun.*, (1993) 333.

20. R. Subramanian *et al.*, *J. Phys. Chem.*, **100** (1996) 16327.

21. P. Paul, Z. Xie, R. Bau, P. D. W. Boyd and C. A. Reed, *J. Am. Chem. Soc.*, **116** (1994) 4145.

22. P. Allemand *et al.*, *Science*, **253** (1991) 301; K. Tanaka *et al.*, *Phys. Rev.*, **B47** (1993) 221, 7554.

23. P. Allemand *et al.*, *J. Am. Chem. Soc.*, **113** (1991) 2780.

24. H. Moriyama, H. Kobayashi, A. Kobayashi and T. Wayanabe, *J. Am. Chem. Soc.*, **115** (1993) 1185.

25. U. Bilow and M. Jansen, *J. Chem. Soc., Chem. Commun.*, (1994) 403.

26. A. Pénicaud, A. Peréz-Benitez, R. Gleason, E. Muñoz and R. Escudero, *J. Am. Chem. Soc.*, **115** (1993) 10392.

27. B. Miller and J. M. Rosamilia, *J. Chem. Soc., Faraday Trans.*, **89** (1993) 273.

28. C. Caron *et al.*, *J. Am. Chem. Soc.*, **115** (1993) 8505.

29. K. M. Kadish, private communication.

30. S. Miki, M. Kitao and K. Fukunishi, *Tetrahedron Lett.*, **37** (1996) 2049.

31. J. Chen, R. Cai, Z. Huang, H. Wu, S. Jiang and Q. Shao, *J. Chem. Soc., Chem. Commun.*, (1995) 1553.

32. D. E. Cliffel and A. J. Bard, *J. Phys. Chem.*, **98** (1994) 8140.
33. L. S. Sunderlin, J. A. Paulino, J. Chow, B. Kahr, D. Ben-Amotz and R. R. Squires, *J. Am. Chem. Soc.*, **113** (1991) 5489.
34. J. W. Bausch *et al.*, *J. Am. Chem. Soc.*, **113** (1991) 3205
35. G. P. Miller, personal communication.
36. H. H. Wang *et al.*, **30** (1991) 2839.
37. D. R. Buffinger, R. P. Ziebarth, V. A. Stenger, C. Recchia, C. H. Pennington, *J. Am. Chem. Soc.*, **115** (1993) 9267.
38. D. W. Murphy *et al.*, *J. Phys. Chem. Solid.*, **53** (1992) 1321.
39. See K. Tanigaki *et al.*, *Nature*, **356** (1992) 419, and references contained therein.
40. A. F. Hebard *et al.*, *Nature*, **350** (1991) 600.
41. A. R. Kortan *et al.*, *Nature*, **355** (1992) 530.

6

Nucleophilic Addition, and Reaction of Fullerene Anions with Electrophiles

The strong electrophilic character of the fullerenes makes them very reactive towards both nucleophilic addition and substitution. Nucleophilic addition has been carried out with both neutral and charged nucleophiles. The species formed in the latter case are described as e.g. 2-hydro[60]fullerene-1-uide (addition of H⁻ to position 2 creating charge at position 1). Anions may also be formed by proton loss from the cage giving e.g. 1,2-dihydro[60]fulleren-1-ide (loss of H⁺ from position 1). Subsequent reaction of the unstable intermediate anions with electrophiles (unreactive towards the neutral species) provides important synthetic routes.

6.1 Reactions with Neutral Nucleophiles

For these to occur requires a very reactive nucleophile; consequently all known examples concern reactions with amines. The overall reaction thus consists of *cis* addition of NR_2 and H (Fig. 6.1), as in the case of the addition of methylamine, ethylamine, propylamine, dodecylamine and morpholine,[1,2] giving water-soluble products. The number of groups added are six for the bulky morpholine, and twelve for propylamine. Nothing is yet known concerning the addition patterns in these reactions, but the twelve-fold addition could be that shown in Fig. 6.2, the conjectured structure having two groups of octahedrally located addends, with a 1,2,3,4-addition pattern (see Sec. 3.1) in six rings; this pattern also creates two hexagons with increased aromaticity due to reduced strain in the three pentagons adjacent to each.

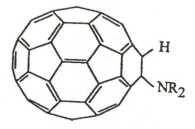

Fig. 6.1 1,2-Addition of HNR$_2$ to [60]fullerene.

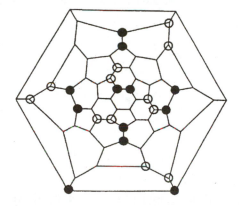

Fig. 6.2 Conjectured octahedrally-located addition sites (filled and open circles) for twelve-fold addition of HNR$_2$ to [60]fullerene.

Both ESR signals and colour changes show that up to three steps can be involved in these additions. A single electron transfer giving an intermediate radical anion is followed by radical combination to give an anion, and finally a proton is added. The mechanism for the addition of *t*-butylamine appears to be different since it only takes place in polar solvents such as dimethylformamide.[1] Tertiary amines, such as triethylamine,[3] tetrakis(dimethylamino)ethene,[4] and 1,8-diazabicyclo[5,4,0]undec-7-ene,[5] lack a hydrogen to complete the normal addition and so transfer an electron to form salt-like products.

Under photochemical irradiation, triethylamine and derivatives bond to [60]fullerene via the carbons adjacent to nitrogen, to give cycloadducts

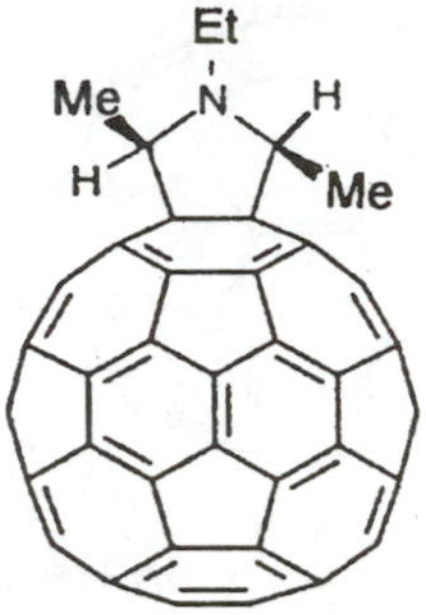

Fig. 6.3 Product of addition of triethylamine to [60]fullerene.

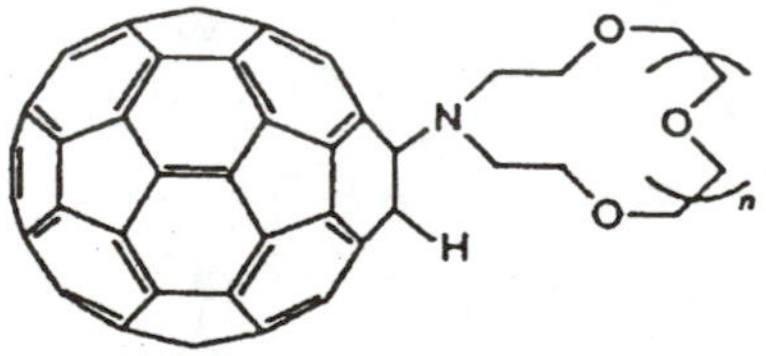

Fig. 6.4 Addition of aza-crown ethers ($n = 2,3,4$) to [60]fullerene (1,2-addition shown).

e.g. *N*-ethyl-*trans*-2′,5′-dimethylpyrrolidino[3′,4′:1,2][60]fullerene (Fig. 6.3).[6] These reactions should be accompanied initially by hydrogen addition at the cage sites adjacent to the positions of bond formation to the addend. However, this introduces double bonds into pentagons which is unfavourable and so both hydrogens are oxidatively eliminated, probably by singlet oxygen. Further elimination of hydrogen and even methylene from the addend can also occur.[7] Pyrrolidino[60]fullerenes can also be obtained simply by reaction of the fullerene with ammonia in the presence of aldehydes, though with phenylacetaldehyde the product is 1-benzyl-1,2-dihydro[60]fullerene instead.[8]

Monoaddition products are obtained from the reaction of aza crown ethers (1-aza-12-crown-4, 1-aza-15-crown-5, and 1-aza-18-crown-6) with [60]fullerene (Fig. 6.4), and appear to be a mixture of both 1,2- and 1,4-isomers;[9] the latter evidently arising from the steric hindrance which somewhat disfavours 1,2-addition (see Sec. 3.1).

1,2-Diamines, e.g. *N,N*′-dimethylethylenediamine, piperazine, or homopiperazine (Fig. 6.5, $R^1, R^2 = $ Me, Me; $(CH_2)_2$; $(CH_2)_3$, respectively

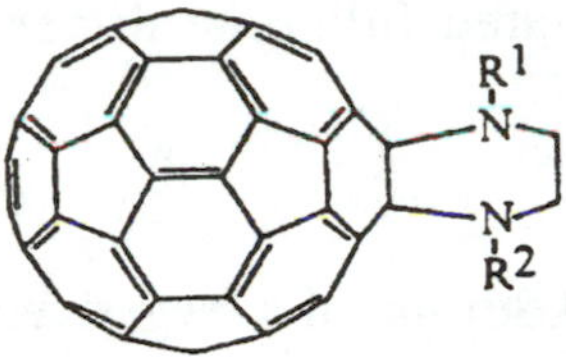

Fig. 6.5 Formation of dehydrogenated 1,2-diamino derivatives of [60]fullerene.

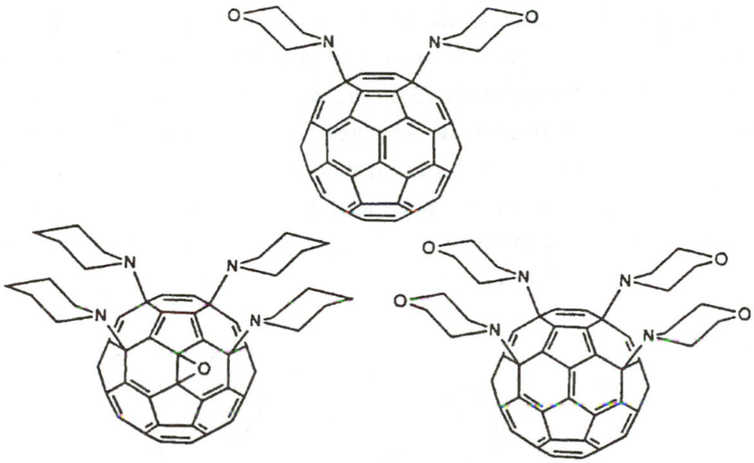

Fig. 6.6 Products of the reaction of morpholine and piperidine with [60]fullerene.

give 1,2-diamino addition products with [60]fullerene in high yields. *Bis* adducts are also detected, and for piperazine addition these have been chromatographically separated into six regioisomers,[10] three of which have been characterised as involving addition across the 1,2:14,15 (*cis*-2),[11] 1,2:18,36 (*e*)[11] and 1,2:51,52 (*trans*-2) bonds.[12] (Note: the numbering in Ref. 11 is incorrect.) Enantiomers of *trans*-1,2-di-*N*-methylaminocyclohexane have been added to [60]fullerene to produce chiral derivatives.[13] The oxidative elimination of hydrogen (see above) gives here an intermediate aminated radical, confirmed by the 1,4-addition of either morpholine or piperazine with [60]fullerene which gives not only the products shown in Fig. 6.6 (notable for the similarity of their regiochemistry to that of hexabromo- and hexachlorofullerenes,

Chap. 7), but also aminated fullerene dimers (resulting from radical combination).[14]

6.2 Anions Formed by Addition of a Negatively Charged Nucleophile

Cyano adducts, $C_{60}(CN)_n^-$, $n = 1,3,5$ with 1 dominant, result from electrospray reaction of [60]fullerene with NaCN;[15] there is evidence also that species consisting of two fullerene cages are also present, a result relevant in view of the results of Komatsu *et al.* on C_{120} (see Chap. 12). Three dianions $C_{60}(CN)_n^{2-}$, $n = 2, 4, 6$, are also produced, and calculations indicate that the structures should differ from those of the neutral species, due to the need to minimise repulsion between the charges. For example, the $n = 6$ dianion structure is predicted to be that shown in Fig. 6.7. Notably, this contains the same pentagonal array of addends found in $C_{60}Br_6$ and $C_{60}R_5^-$ (Chap. 7) but is here especially favoured by the aromaticity created in the (6π electron) central pentagon.

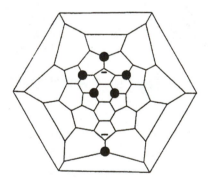

Fig. 6.7 MOPAC/AM1 predicted structure for $C_{60}(CN)_6^-$ ($\bullet$ = CN).

Reaction of lithium 9-fluorenide with [60]fullerene in THF in the presence of air results in 1,4-attachment of two 9-fluorenyl groups to give **6.1**, presumed to occur *via* one-electron oxidation of the first-formed anion, followed by addition of the second anion and further oxidation. The structural assignment was confirmed by the preparation of the corresponding derivative **6.2** in which

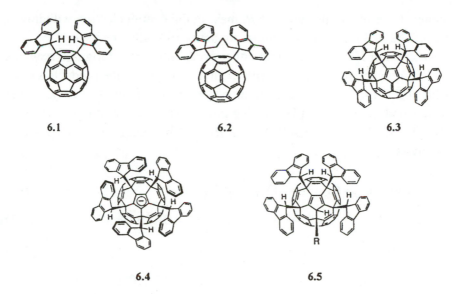

6.1   6.2   6.3

6.4   6.5

the 9-hydrogens of the fluorenyl addends are linked with a trimethylene group.[16] Similar use of potassium 9-fluorenide results in the 1,4-addition of four 9-fluorenyl groups to give the tetrakis adduct **6.3**, which can react with a further potassium 9-fluorenide to give **6.4** which also has a stabilising 6π pentagonal ring. Addition of acid to **6.4** gives **6.5** (R = 9-fluorenyl), and related derivatives (**6.5**, R = CN, C $\equiv$ C-hexyl) compounds are obtainable by reacting LiR with **6.3**.[17]

Reaction of α-halocarbanions with fullerenes (Bingel reaction) forms the first step in one route to the preparation of preparation of methanofullerenes, described in Sec. 9.1.1.3

6.3 Reaction of Anions with Electrophiles

More controlled reaction of [60]fullerene with NaCN followed by quenching of the $C_{60}CN^-$ intermediate with appropriate reagents has permitted isolation of derivatives $RC_{60}CN$ (R = H, CN, Me, 4-t-butylbenzyl). Addition to the cage usually makes it less electron withdrawing (see Sec. 4.2), and thus for example the first reduction potential of $C_{60}F_{48}$ is at 0.79 v compared to -0.56 v for C_{60} itself. By means of cyclic voltammetry it was shown that the CN addend counters the effect of the addends R, making the overall electron withdrawal by the cage in the derivatives similar to that of [60]fullerene itself.[18]

Reaction of [60]fullerene with either organolithium or Grignard reagents in THF, followed by quenching with methyl iodide gives the products shown in Eqs. 6.1 – 6.4.[19] A notable feature of these results is the tendency (as yet unexplained) to add multiples of 10 groups, as found also in the reaction of n-butyllithium with [60]fullerene followed by reaction with CO_2 and hydrolysis.[20]

$$PhMgBr(xs.) + C_{60} \rightarrow C_{60}Ph_{10}Me_{10} \qquad (6.1)$$

$$PhLi + C_{60} \rightarrow C_{60}Ph_xMe_y \ (x = 0 - 10; y = 10 - 1) \qquad (6.2)$$

$$t\text{-}BuMgBr + C_{60} \rightarrow C_{60}t\text{-}Bu_{10}Me_{10} \qquad (6.3)$$

$$t\text{-}BuLi + C_{60} \rightarrow C_{60}t\text{-}Bu_xMe_y \ (x = 2, 1, 0; y = 2, 1) \qquad (6.4)$$

Under more controlled conditions (using either t-BuLi or EtMgBr), $C_{60}t$-BuH and C_{60}EtH can be obtained as well as other derivatives believed to be formed *via* a $C_{60}(t\text{-}Bu)_n{}^{n-}$ intermediate.[21,22] The bulk of the t-butyl group makes the formation of 1,4-C_{60}Ht-Bu not wholly unfavourable compared to the 1,2-isomer and so both are formed, the 1,4-isomer converting slowly to the 1,2-isomer on standing. However, more recent work (see Sec. 6.4) shows that in the absence of base the 1,4-isomer is stable indefinitely. Electrochemically, C_{60}Ht-Bu is 0.15 v harder to reduce than [60]fullerene,[22] due to the electron supply to the cage by the addends. The high electron withdrawal by the cage makes C_{60}Ht-Bu a very strong carbon acid, having a pK_a of 5.7, with a weak (71 kcal mol^{-1}) C-H bond dissociation energy (see also Sec. 4.2.1).

Reaction of lithium 9-fluorenide with [60]fullerene in THF (see above) and subsequent reaction of the intermediate carbanion with acid (proton electrophile) gives 1-(9-fluorenyl)-1,2-dihydro[60]fullerene.[15] Other diaddended compounds C_{60}HX (X = octyl, but-3-enyl, 1,3-dioxalan-2-yl, prop-2-yl, phenyl, CH_2Ph, tributylstannyl, C ≡ CH, C ≡ CPh, C ≡ CSiMe$_3$, CH_2SiMe_2R and SiR$_3$ etc.) have been made in this way, as well as compounds C_{70}HX (X = Ph, Me).[23-28] The latter compound has C_s symmetry showing that addition has occurred across the 1,2-bond, so the hydrogen is probably attached to C-2, i.e. the polar pentagon, because the ^{1}H NMR shift δ of 6.04 is very close to that for the corresponding [60]fullerene derivative. For 1,2-dihydro[70]fullerene the ^1H NMR shift for H-2 is *ca.* 0.9 ppm downfield relative to that for H-1, and values for [70]fullerene are generally *ca.* 1 ppm upfield relative to those for [60]fullerene, so the δ value of 6.04 would seem to be too downfield to be assignable to H-1. The C-Si bond in $C_{60}H.C \equiv CSiMe_3$ is exceptionally resistant to base cleavage, which may arise from repulsive interactions between the (virtually) orthogonal p-orbitals of the cage and those of the addend, resulting in increased pπ-dπ bonding between the triple bond and silicon. This conjecture is supported by the refusal of $HC \equiv C^-$ to react with [60]fullerene, whereas $Me_3SiC \equiv C^-$ does react.[28]

Di-addition has also been obtained yielding compounds $C_{60}SiMe_3Y$ (Y = benzyl, OTs, CHO and CH_2OH)[28] and 1,4-$C_{60}(CH_2SiMe_2X)_2$, the addends in the latter being confirmed as in a 1,4-relationship,[26] attributable to their bulk.

1,4-Addition of benzyl groups (*cf.* Sec. 5.6) also occurs on reaction of benzyl bromide with a dianion complex $C_{60}Fe(CO)_3^{2-}$ (though the mechanism may involve radical anion intermediates).[29] Reaction of [60]fullerene with PhMgBr/ $CuBr.SMe_2$ followed by quenching with NH_4Cl produces $C_{60}Ph_5H$ (see also Sec. 8.6.1), probably through sequential addition by Ph_2Cu^-.[30]

Metal complexes (M = Re, Fe, Ru, and Mn) bound to the 2-position of [60]fullerene through alkyl bridges result from reaction of the anion $C_{60}H^-$ (produced by reaction of [60]fullerene with $LiBH_3H$) with various cationic complexes such as $[(OC)_5Re(\eta^2-C_2H_4)]^+$.[31]

The Reformatsky reaction ($Zn/BrCH_2CO_2Et$) works with [60]fullerene to give 1,2-addition of H and CH_2CO_2Et, though the reaction is accompanied here not only by 1,4-di-addition of two CH_2CO_2Et groups, but also by [1 + 2]addition of $HCCO_2Et$ across a 6,6-bond.[32]

1,2-Addition of hydrogen and phosphorus can be achieved by reaction of [60]fullerene with $R'R''P^-Li^+.BH_3$ to give first a phosphinite-borane, and then by quenching with HCl and removal of the BH_3 group, a fullerene phosphine, $C_{60}HPR'R''$ (R',R'' = Ph; R' = Ph, R'' = Me; R' = Ph, R'' = O-menthyl).[33]

Nucleophilic addition of OH groups seems to occur in reaction of either [60]- or [70]fullerenes with xs. KOH under vacuum;[34] presumably H becomes attached to the cage also, but may then be rapidly oxidised up to OH groups, this being known to readily occur in fullerenols.[35] Approximately 24 hydroxy groups are added to [60]fullerene in reaction with aq. NaOH in the presence of t-butylammonium hydroxide,[36] and the location of the addends is probably the same as in $C_{60}Br_{24}$ (Chap. 7), especially as this would provide an excellent opportunity for global hydrogen bonding.

In the reaction between NaOMe/MeOH and either [60]- or [70]fullerene, species such as $C_{60}(OMe)_n^-$, $n = 1,3,5$ or 7, together with products arising from proton abstraction for the solvent, have been detected by negative ion mass spectrometry, though no isolable products are obtained.[37] The reaction between $PhCH_2ONa$-$PhCH_2OH$ with [60]fullerene in the presence of air produces an isolable dioxalane derivative, via intermediate formation of $PhCH_2OC_{60}^-$ which is presumed to then react with oxygen (which is necessary for the reaction to take place).[38]

6.4 Anions Formed by Proton Loss

Removal of a proton with a strong base, e.g. KOt-Bu, and then reaction of the intermediate fullerenide anion with alkyl halides leads to various alkyl-substituted dihydro[60]fullerenes, the location of the addends depending upon the steric requirements of the addends: if these are not very large then 1,2-addition occurs, otherwise addition is 1,4. In this way compounds containing addend combinations of octyl and 2-R (R = Me, Et, PhCO and tropylium) or t-Bu and tropylium, have been prepared.[39] Base-catalysed proton loss from 1-t-butyl-1,2-dihydro[60]fullerene has been used in the first fullerene kinetic study, which shows that the 1,4-isomer (which contains a double bond in a pentagon) rearranges in the presence of pyridine base to the 1,2-isomer (which does not), with an activation energy of 56 kJ mol^{-1}.[40]

References

1. A. Hirsch, Q. Li and F. Wudl, *Angew. Chem. Intl. Edn. Engl.*, **30** (1991) 1308.
2. R. Seshradi, A. Govindaraj, R. Hagarajan, T. Pradeep and C. N. R. Rao, *Tetrahedron Lett.*, (1992) 2069.
3. J. Pola *et al.*, *Fullerene Sci. & Technol.*, **3** (1995) 229.
4. P. M. Allemand *et al.*, *Science*, **253** (1991) 301; P.W. Stephens *et al.*, *Nature*, **355** (1992) 331.

5. A. Skiebe, A. Hirsch, H. Klos and B. Gotschy, *Chem. Phys. Lett.*, **220** (1994) 138.
6. G. E. Lawson, A. Kitaygorodskiy, B. Ma, C. E. Bunker and Y.-P. Sun, *J. Chem. Soc., Chem. Commun.*, (1995) 2225; K. Liou and C. Cheng, *Chem. Commun.*, (1996) 1423.
7. A. D. Darwish and R. Taylor, unpublished work.
8. A. Komori, M. Kubata, T. Ishida, H. Niwa and T. Nogami, *Tetrahedron Lett.*, **37** (1996) 4031.
9. S. N. Davey, D. A. Leigh, A. E. Moody, L. W. Tetler and F. A. Wade, *J. Chem. Soc., Chem. Commun.*, (1994) 397.
10. K. Kampe, N. Egger and M. Vogel, *Angew. Chem. Intl. Edn. Engl.*, **32** (1993) 1174.
11. A. L. Balch, A. S. Ginwalla, M. M. Olmstead and R. Herbst-Irmer, *Tetrahedron*, **52** (1996) 5021.
12. A. L. Balch, B. Cullison, W. R. Fawcett, A. S. Ginwalla, M. M. Olmstead and K. Winkler, *J. Chem. Soc., Chem. Commun.*, (1995) 2287.
13. M. Maggini, G. Scorrano, A. Bianco, C. Toniolo and M. Prato, *Tetrahedron Lett.*, **36** (1955) 2845.
14. G. Schick, K.-D. Kampe and A. Hirsch, *J. Chem. Soc., Chem. Commun.*, (1995) 2023.
15. G. Khairallah and J. B. Peel, *Chem. Commun.*, (1997) 253; *Chem. Phys. Lett.*, **268** (1997) 218.
16. Y. Murata, K. Komatsu and T. S. M. Wan, *Tetrahedron Lett.*, **37** (1996) 7061.
17. Y. Murata, M. Shiro and K. Komatsu, *J. Am. Chem. Soc.*, **119** (1997) 8117.
18. M. Kershavarz, B. Knight, G. Srdanov and F. Wudl, *J. Am. Chem. Soc.*, **117** (1995) 11371.
19. F. Wudl in *Buckminsterfullerenes*, Eds. W. E. Billups and M. A. Ciufolini, VCH, (1993) 317.
20. P. Ya. Bayushkin, G. A. Domrachev, V. L. Karnatsevich and R. Taylor, unpublished work.
21. A. Hirsch, A. Soi and H. S. Karfunkel, *Angew. Chem. Intl. Edn. Engl.*, **31** (1992) 766.
22. P. J. Fagan, P. J. Krusic, D. H. Evans, S. A. Lerke and E. Johnston, *J. Am. Chem. Soc.*, **114** (1992) 9697.

23. A. Hirsch, T. Grösser, A Skiebe and A. Soi, *Chem. Ber.*, **126** (1993) 1061.
24. K. Komatsu, Y. Murata, N. Takimoto, S. Mori, N. Sugita and T. S. M. Wan, *J. Org. Chem.*, **59** (1994) 6101; K. Komatsu, N. Takimoto, Y. Murata, T. S. M. Wan and T. Wong, *Tetrahedron Lett.*, **37** (1996) 6153.
25. H. L. Anderson, R. Faust, Y. Rubin and F. Diederich, *Angew. Chem. Intl. Edn. Engl.*, **33** (1994) 1366.
26. H. Nagashima, H. Terasaki, E. Kimura, K. Nakajima and K. Itoh, *J. Org. Chem.*, **59** (1994) 1246; H. Nagashima, H. Terasaki, Y. Saito, K. Jinno and K. Itoh, *J. Org. Chem.*, **60** (1995) 4966.
27. T. Kusukawa and W. Ando, *Angew. Chem. Intl. Edn. Engl.*, **35** (1996) 1315.
28. P. Timmerman *et al.*, *Tetrahedron*, **52** (1996) 4925; P. Timmermann, L. E. Witschel, F. Diederich, C. Boudon, J. Gisselbtracht and M. Gross, *Helv. Chim. Acta*, **79** (1996) 6.
29. S. Miki, M. Kitao and K. Fukunishi, *Tetrahedron Lett.*, **37** (1996) 2049.
30. M. Sawamura, H. Iikura and E. Nakamura, *J. Am. Chem. Soc.*, **118** (1996) 12850.
31. W. Beck, H.-J. Bentele and S. Hüffer, *Chem. Ber.*, **128** (1995) 1059.
32. G.-W. Wang, Y. Murata, K. Komatsu and T. S. M. Wan, *Chem. Commun.*, (1996) 2059.
33. S. Yamago, M. Yanagawa and E. Nakamura, *J. Chem. Soc., Chem. Commun.*, (1994) 2093.
34. A. Naim and P. B. Shevlin, *Tetrahedron Lett.*, **33** (1992) 7097.
35. N. S. Schneider, A. D. Darwish, H. W. Kroto, R. Taylor and D. M. Walton, *J. Chem. Soc., Chem. Commun.*, (1994) 463.
36. J. Li, A. Takeuchi, M. Ozawa, X. Li, K. Saigo and K. Kitazawa, *J. Chem. Soc., Chem. Commun.*, (1993) 1784.
37. S. R. Wilson and Y. Wu, *J. Am. Chem. Soc.*, **115** (1993) 10334.
38. G.-W. Wang, L.-H. Shu, S.-H. Wu, H.-M. Wu and X.-F. Lao, *J. Chem. Soc., Chem. Commun.*, (1995) 1071.
39. H. Okamura, Y. Murata, M. Minoda, K. Komatsu, T. Miyamoto and T. S. N. Wan, *J. Org. Chem.*, **61** (1996) 8500; Y. Murata, K. Motoyama, K. Komatsu and T. S. M. Wan, *Tetrahedron*, **52** (1996) 5077; T. Kitagawa, T. Tanaka, Y. Takata, K. Takeuchi and K. Komatsu, *Tetrahedron*, **53** (1997) 9965.
40. F. Banim, D. J. Cardin and P. Heath, *Chem. Commun.*, (1997) 25.

7

Radical Reactions

The electron-withdrawing properties of fullerenes make them reactive towards radicals. Two main classes of reactions have been studied, *viz.* halogenation, which leads to products amenable for further reactions, and addition of (mainly) carbon radicals. The latter reactions have been the subject of extensive ESR studies, but with improved HPLC separations becoming available, may lead in the future to isolation of numerous new derivatives.

7.1 Halogenation

Fluorination of fullerenes has been confirmed as a radical reaction, as must also be chlorination and bromination since these reactions are not accelerated by Lewis acid catalysts showing that electrophilic halogenation does not take place. Apart from producing synthons for further fullerene derivatisation, halogenation is especially interesting because of the different additions patterns produced by the different sizes of the halogens. This facilitates understanding of the mechanisms that govern addition to fullerenes. Thus bromination and chlorination will add a maximum of 24 halogens, whereas fluorination will readily add 48 fluorines, and more under forcing conditions.

The halogenofullerenes are not very thermally stable and eliminate halogen on heating to *ca.* 150°C, the stability order being fluoro- > chloro- > bromo- > iodofullerenes (the latter in fact are too unstable to exist). In the case of the bromo[60]fullerenes, the decomposition is more rapid, the fewer bromines that are present.[1] This behaviour (which is paralleled in hydrogenation) may reflect the shorter pathway required to return to the [60]fullerene ground state, or reduced strain due to the presence of a larger number of sp^3 carbons in the cage. The solubility order is fluoro- > chloro- > bromofullerenes and indeed the bromofullerenes are difficult to use in syntheses because of their low solubility. The halogenofullerenes are much more reactive than the corresponding halogenoalkanes, and are susceptible towards nucleophilic substitution by atmospheric water, but this is sufficiently slow to make dry box conditions unnecessary for most manipulations. Decomposition is faster if the compounds are stored in solution.

7.1.1 *Fluorination*

Fluorination of fullerenes was thought to be a promising route to the formation of super lubricants ('superTeflon') with $C_{60}F_{60}$ at the forefront of thinking.[2] However this overlooked the fact that unlike the carbon backbone in Teflon (which twists to eliminate eclipsing interactions, and also completely shields the carbons from attack by other reagents), the fullerene cages cannot do this to any significant extent. It is therefore difficult to add more than 48 fluorines to the [60]fullerene (though it has been achieved on a small scale in fluorination by fluorine gas),[3,4] and moreover the fluorofullerenes are very reactive towards nucleophilic substitution.[5] Up to *ca.* 76 fluorines can be added to [60]fullerene under extreme conditions (either fluorination with UV irradiation,[6] or reaction with KrF_2 in anhydrous HF[7]) and here cage opening must occur.

Two main conditions of fluorination have been used, *viz.*, fluorination with fluorine gas, and with metal fluorides. The former condition is difficult to control, and leads to a wide range of products, whereas the latter is easier to control, but is more expensive and ultimately unsuitable for large scale production. Fluorination with KrF_2, XeF_2, BrF_5 and IF_5 have been used in limited experiments.

7.1.1.1 *Fluorination with Fluorine Gas*

The difficulty with this technique is that no solvent that is resistant to fluorine will dissolve fullerenes, consequently fluorination has to be carried out under heterogeneous conditions. The lattice packing of pure fullerenes, especially [60]fullerene, is so close that fluorine cannot penetrate readily. As fluorination proceeds the material swells as the outer cages become increased in size through addition, and expand away from the lattice so allowing fluorine to penetrate to the next layer. Consequently no fluorination of the inner layers occurs until that of the outer layers is substantially complete, with each molecule in the outer layer becoming highly fluorinated. Thus, if [60]fullerene is treated with fluorine gas at 70°C for 24 h, the product consists of a mixture of highly fluorinated- and unreacted fullerene, the latter being recoverable in its original state.[8] Differential packing in the solid is also the reason why fluorination of pure[60]fullerene[9] is much slower than that of a [60/70]fullerene mixture;[10] comparable close packing cannot be achieved in the latter.[11,12] The packing in [70]fullerene is also not as good as in [60]fullerene, consequently it fluorinates faster than the latter,[8] this being the reverse reactivity order to that observed in most reactions.

The level of fluorination that can be achieved, and the time taken for it to occur, depends upon the temperature of fluorination and (as noted above) the purity of the fullerene.[3,6,10,12–16] As a typical example, fluorination of pure [60]fullerene at 70°C requires *ca.* 6 weeks for completion and produces a cream-white solid, and a pink one being obtained (more rapidly) from [70]fullerene.[8] The colour changes that accompany the fluorination are associated with gradual penetration of the fullerene lattice. The mass spectra showed that a wide range of fluorinated species are produced, as indicated by the broad bands in the 1100–1175 cm^{-1} region; the maximum levels of fluorination that can be achieved in significant concentration are around $C_{60}F_{48}$ and $C_{70}F_{56}$ with various research groups finding the species of maximum *concentration* being a little lower than these.

The products of these fluorinations usually contain oxides[3,6,8,12] which may derive from various sources, such as moisture occluded in the surface of the reaction vessel,[6] moisture in the air or solvents during manipulations after fluorination, and oxygen trapped in the fullerene lattice.[8] The oxide concentrations are increased immediately on dissolving fluorofullerenes in moist solvents, whence a whole range of derivatives containing up to eighteen oxygens are obtained.[8,14] These are presumed to be in the form of epoxides and derived from nucleophilic substitution of F by OH, followed by elimination of HF. In the absence of a Lewes acid catalyst topolarise the C-F bond (thereby permitting a front-side direct nucleophilic replacement), the substitution mechanism may involve addition-elimination (Fig. 7.1).[8]

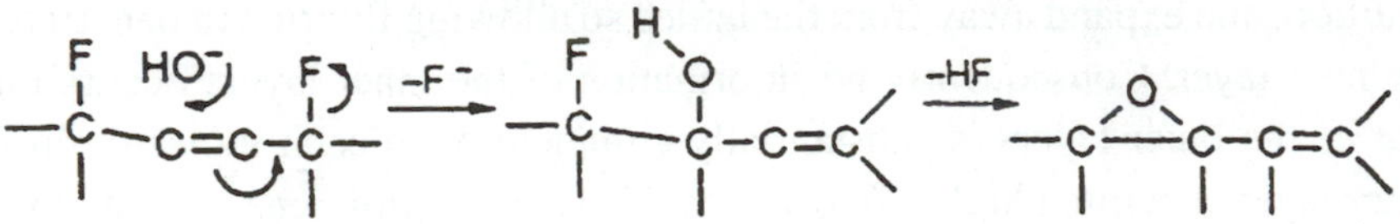

Fig. 7.1 Possible addition-elimination mechanism leading to the formation of epoxides from fluorofullerenes.

One very notable feature of the reaction of fluorinated [60]fullerene with moist methanol was the increase in intensities of species containing 18 and 36 fluorine atoms, which must therefore have enhanced stability;[8] a similar conclusion may be deduced from other fluorination experiments.[3] This indicated that fluorination paralleled hydrogenation, since both $C_{60}H_{18}$ and $C_{60}H_{36}$ had already been detected in hydrogenation (Chap. 4). Subsequent experiments (Sec. 7.1.1.2) have confirmed this prediction. A curious but unexplained feature arising from the addition of moist methanol to fluorinated [60]fullerene is the formation of trifluoromethyl[60]fullerenes [which are much the most volatile of all fullerene compounds (see Sec. 7.2.1)]. Fluorofullerenes themselves are also volatile, and this volatility increases with increasing degree of fluorine addition. This feature is utilised in fluorination with metal fluorides (Sec. 7.1.1.2).

Another and as yet unexplained property of the reaction products is that the ^{19}F NMR spectrum shows a broad hump (which would be expected in view of the mixture of products) but centred on this, at around -152 ppm, is a sharp singlet. This evidently is due to a symmetrical species (of which $C_{60}F_{60}$ was originally the prime suspect)[9] but more recent work indicates that this cannot be the cause since the concentration is too low.[4,15] It is now known to arise from an (unknown) reaction with the solvent, as has been demonstrated in the measurement of the NMR spectrum of pure $C_{60}F_{48}$. This changes dramatically on being run in either diethyl ether or THF, whence this peak appears and all others are lost.[16] Fluorinated [70]fullerene also shows a series of sharp singlets in the ^{19}F NMR, for reasons as yet unknown.[8]

Fluorination at higher temperatures leads to a product with a smaller range of fluorinated species, and with a higher overall content. This has been

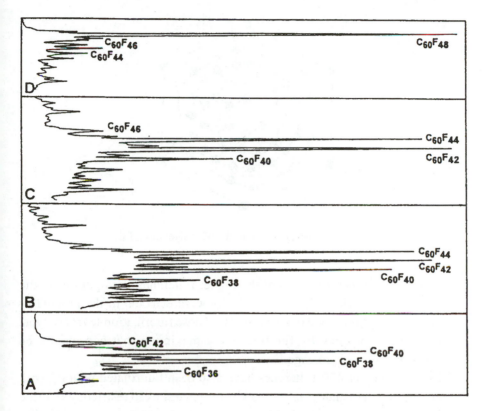

Fig. 7.2 EI mass spectra for the products of fluorination of [60]fullerene at: (A) 4 h at 70°C; (B) plus 5 h at 220°C; (C) plus 11 h at 275°C; (D) plus 3 h at 275°C.

shown by a reaction time/temperature product mass spectrometry study, giving the result shown in Fig. 7.2. Under forcing conditions 315°C, 16 h it is thus possible to prepare pure $C_{60}F_{48}$,[16] which may also be prepared using a longer reaction time, lower temperature, and the addition of sodium fluoride to the reaction mixture.[17] This, the first fluorofullerene to be characterised, crystallises from chloroform as large colourless prisms but immediately forms coloured charge transfer complexes with electron-rich aromatic solvents such as toluene.[16] The ^{19}F NMR spectrum indicates the structure to consist of an RR and SS enantiomeric pair[17] (Fig. 7.3), and the IR spectrum shows detailed structure,[16] in contrast to that obtained with fluorofullerene mixtures.

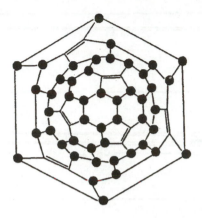

Fig. 7.3 Schlegel diagram of the SS isomer of $C_{60}F_{48}$.

Due to the presence of traces of the higher fullerenes as impurities in the starting material, $C_{70}F_{56}$, $C_{78}F_{64}$ and $C_{84}F_{62}$ were detected in the product of fluorination of [60]fullerene by F_2 at 315°C. These fluorination levels suggest that 80% site occupancy by fluorine is the maximum that can readily be accommodated on a given cage.[16]

Bromo- and chloro[60]fullerenes have also been fluorinated with F_2, with the aim of using these halogens to block some cage sites so restricting fluorine addition.[18] Since bromine and chlorine are lost from the cage more readily than fluorine on heating, this could produce a fluorinated fullerene of lower fluorine content than could be obtained by direct fluoination of the parent fullerene. The product of $C_{60}Br_{24}$ fluorination does indeed have a lower fluorine content which is reduced further on heating strongly. Fluorination of $C_{60}Br_6$ is less successful due possibly to the presence in the bromofullerene lattice of benzene. This becomes smoothly fluorinated to hexafluorobenzene — a novel way to control direct fluorination of aromatics! The EI mass spectrum of fluorinated $C_{60}Br_8$ is particularly interesting in that $C_{60}F_{18}$ and both its mono- and di-oxide are the most abundant species produced, the fragmentation pattern being also similar to that obtained with $C_{60}F_{18}$ itself.

Fluorinations of $C_{60}Cl_6$ and $C_{60}Cl_{24}$ with XeF_2 were relatively unsuccessful, but fluorination of $C_{60}Cl_{24}$ with IF_5 produces $C_{60}Cl_{18}F_{14}$.[18] Two interesting aspects arise here. Since the maximum number of chlorines that can be readily accommodated on the [60]fullerene cage is 24, whilst that of fluorine is 48, this indicates that the latter has half the space requirement of the former. A compound with 18 chlorines should thus be able to accommodate 12 fluorines, close to the number observed. A second observation concerns the fluorination mechanism. If the reaction is carried out in carbon tetrachloride, CCl_3F is detected as a by-product, and this can have been produced only by a *radical* fluorinating species.[18]

7.1.1.2 *Fluorination with Metal Fluorides*

This technique makes use of the volatility of the fluorofullerenes relative to the parent compounds. By heating a mixture of an appropriate metal fluoride and fullerene under vacuum, the fluorofullerenes are swept away from the mass once a certain fluorination level is achieved, and so reaction ceases. The products are deposited in a cold zone, the material furthest removed from the reaction zone being the more volatile and the least coloured. (This fractional sublimation property could in principle be also used to separate out the products of fluorination by fluorine gas.) A lower fluorination level is thus achievable by using a higher temperature (at which the less-fluorinated fullerenes acquire sufficient volatility) but this in turn requires a metal fluoride that releases fluorine only at a higher temperature. A reactivity series for metal fluorides is $CoF_3 > MnF_3 > FeF_3$.[19]

Use of MnF_3 produces pure $C_{60}F_{36}$ which consists of two (white/off-white) isomers, one of which has T symmetry (Fig. 3.9, • = F), and the other has C_3 symmetry, believed to be that shown isomer shown in Fig. 3.10, • = F.[20] These isomers are obtained in *ca.* 1:3 ratio, respectively; this isomer number and ratio parallels that found for hydrogenation (Sec. 4.2.5) which demonstrates the similarity between hydrogenation and fluorination.

Fluorination of [60]fullerene by K_2PtF_6 produces (yellow-green) $C_{60}F_{18}$ (*cf.* Fig. 3.8, • = F),[21] which is isostructural with $C_{60}H_{18}$, and has enhanced stability due to the presence of a fully delocalisable benzenoid ring. (Delocalisable because strain is reduced by the adjacency of the sp^3-hybridised carbons.) This compound is believed to be an intermediate on the pathway to the formation of T and C_3 $C_{60}F_{36}$ [the motif can be seen in their respective Schlegel diagrams (Figs. 3.9 and 3.10)] and, moreover, it is produced on thermal decomposition of $C_{60}F_{36}$. $C_{60}F_{18}$ readily undergoes nucleophilic substitution into aromatics (Sec. 8.6.3), and is very polar because the fluorines are all located at one end of the molecule. Derivatives $C_{60}F_{18}O$ and $C_{60}F_{18}CF_2$ have also been isolated, and each case the additional O and CF_2 are located on one of the symmetry planes.[22] The latter addend also shows a similarity with hydrogenation, which is often accompanied by the formation of methylene adducts (Sec. 4.2.7). This is also consistent with a study involving C_{1s} XPS and solid state NMR which showed that at increasing fluorination levels, C-C bond-breaking accompanies the reaction, with CF_2 and CF_3 groups being produced.[23]

A further similarity to hydrogenation is also found in fluorination of [70]fullerene by MnF_3.[19] This gives a mixture of $C_{70}F_{36/38/40}$, with $C_{70}F_{38}$ the main product, exactly the same pattern as in hydrogenation. Theoretical proposals as to the nature of these products have been proposed, but no definitive details of the structure are yet available;[24] many isomers of $C_{70}F_{36}$ and the other addition levels are present.[25]

A preliminary study of the fluorination of [76]- and [84]fullerene by K_2PtF_6 indicates that the main product are respectively $C_{76}F_{38}$ and $C_{84}F_{40}$; for the latter exactly the same number of addends are present as in the stable hydrogenation product (Sec. 4.2.8), but nothing is yet known concerning the addend locations.[25]

7.1.1.3 *Fluorination with Krypton Difluoride*

In contrast to fluorination with F_2 gas, fluorination with KrF_2 in HF produces considerably lower concentrations of oxygenated species in the mass spectrum.[7] Whilst this suggests that these species originate in the former work largely from the fluorine used, the difference could arise from the mass spectrometric technique employed; solid samples were analysed in this work, whereas solutions in dichloromethane (DCI probe) were used in the studies giving high concentrations of oxygenated species.

This fluorination condition leads to substantial cage rupture, with formation of species (which must be cage opened) up to $C_{60}F_{78}$. This number of fluorines is almost exactly that observed in hyperfluorination using UV irradiation,[6] which suggests that beyond this fluorination level, substantial cage degradation to smaller fragments occurs.

7.1.1.4 *Properties of Fluorofullerenes*

The fluorofullerenes are very soluble in polar solvents and significantly soluble even in e.g. hexane, which allows them to be processed by HPLC, and to the extent of separation of isomers. As might be expected, the electron affinities of the highly fluorinated fullerenes are much higher (*ca.* 1.5 eV) than those of the parent fullerenes,[26] whilst the affinities of $C_{60}F_2$ and $C_{70}F_2$ are *ca.* 0.1 eV higher than those of the parents.[27] Together these results suggest that addition of each fluorine to the cage raises the affinity by about 0.05 eV, with a gradually decreasing effect as more fluorines are added. The electron affinity of the higher fluorinated species is such that doubly charged anions, e.g. $C_{60}F_{48}^{2-}$, can be readily generated in the mass spectrometer.[28] Fluorofullerenes are also oxidising agents,[29] and they can fluorinate aromatics due to slow liberation of fluorine on heating;[29,30] direct fluorination of aromatics is otherwise generally difficult.[31]

Fluorofullerenes readily form coloured charge transfer complexes with aromatics, e.g. with toluene the colour intensity deepens and is red-shifted, the larger the number of fluorines on the cage.

7.1.2 *Chlorination*

Chlorination of [60]fullerene by chlorine gas at 250°C gives an orange solvent-soluble product, indicated by chlorine uptake to be $C_{60}Cl_{24}$.[32] This eliminates chlorine on heating at 400°C, precluding mass spectroscopic confirmation of the chlorine content, but it is highly probable that the compound is isostructural with $C_{60}Br_{24}$ (Sec. 7.1.3). In general, EI mass spectra of polychlorofullerenes cannot be obtained due to ready elimination of chlorine, but is possible in e.g. $C_{60}Ph_5Cl$ (Sec. 8.6.1) when there is no suitable elimination partner for the chlorine. More recently a FAB mass spectrum of $C_{60}Cl_{24}$ has been successfully obtained, the only known example of a mass spectrum of a polychlorinated fullerene.[18] Reaction of [60]fullerene with liquid chlorine at −35°C produces a brown solid of approximate composition $C_{60}Cl_{12}$ and this dechlorinates on heating at 200–350°C, i.e. at a lower temperature than the more highly chlorinated species (above).[33] This follows the general pattern of fullerene derivative stability, namely that this is higher for compounds having a larger number of addends.

The most important chloro compound yet made from [60]fullerene is $C_{60}Cl_6$, (1,2,4,11,15,30-hexachloro-1,2,4,11,15,30-hexahydro[60]fullerene) produced by reaction of [60]fullerene with ICl in benzene.[34] Its importance derives from the fact that it is soluble in many solvents (thereby facilitating subsequent reaction) and many derivatives may be made from it. The chlorination is slower if carried out in toluene, confirming it to be a radical reaction, since toluene is a radical scavenger. Fig. 7.4(a) shows the structure of $C_{60}Cl_6$ which is interesting from a number of aspects. It is isostructural with $C_{60}Br_6$ (Sec. 7.1.3), and like this derivative is evidently produced by a sequence of 1,4-additions (preferred to 1,2-additions because of the low steric requirement), even though this process requires unfavourable location of a double bond in a pentagon (see also Sec. 3.4). It is thus the kinetic product, but it appears also to be the thermodynamic product, given that calculations of the most stable structure of $C_{60}H_6$ (Sec. 4.2.3) produce the same regiochemical pattern. A by-product of the above reaction is the formation of $C_{60}Cl_{12}$ (detected by its conversion into $C_{60}Ph_{12}$ the structure of which has yet to be determined).

It might be supposed that chorination of [70]fullerene with ICl would give $C_{70}Cl_6$ with the chlorines arranged around one of the poles (which have the same structure as [60]fullerene. This however is not the case, as described in Sec. 3.4, and so $C_{70}Cl_{10}$ (7,8,19,26,33,37,45,49,53,63-decachloro-7,8,19,26,33,37,45,49,53,63-decahydro-[60]fullerene) is produced instead as shown in Fig. 7.4(b).[35]

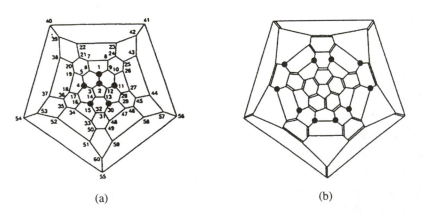

(a)   (b)

Fig. 7.4 Schlegel diagrams (• = halogen) of (a) $C_{60}Cl_6$ and $C_{60}Br_6$; (b) $C_{70}Cl_{10}$.

7.1.3 *Bromination*

Bromination of fullerenes requires either concentrated solutions of bromine or neat bromine for reaction to occur, and this provides further indication that a radical reaction is involved. An early report[32] that either 2 or 4 bromines could be added to [60]fullerene was incorrect. Bromination with neat bromine produces a yellow-orange microcrystalline solid having the approximate composition $C_{60}Br_{24-28}$.[33] Since twenty-eight bromines cannot be attached to [60]fullerene in a symmetrical way, it suggested that the derivative was $C_{60}Br_{24}$ with additional bromine trapped in the interstices of the crystal. This was confirmed by single crystal X-ray studies, which revealed the T_h structure shown in Fig. 7.5; the corresponding Schlegel diagram is shown in Fig. 3.11. The structure can be envisaged as consisting either of twelve hexagons, each with a 1,4-arrangement of two bromine atoms (in a boat conformation), or of eight hexagons each with a 1,3,5-arrangement of three bromine atoms (in a chair conformation). Although each pentagon contains a double bond, it also contains two sp^3-hybridised carbons thereby diminishing the strain that would otherwise be present. $C_{60}Br_{24}$ (1,4,7,10,12,14,16,19,22,24,27,29,31,33,36,38,41,43,46,49,52,54,57, 60-tetracosabromo-1,4,7,10,12,14,16,19,22,24,27,29,31,33,36,38,41,43,46, 49,52,54,57,60-tetra-cosahydro[60]fullerene) is very insoluble, and therefore has not been used in derivatisation.

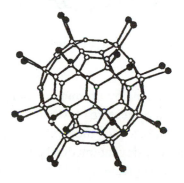

Fig. 7.5 Structure of $C_{60}Br_{24}$.

Bromination of [70]fullerene under the same conditions also gives a product of approximate composition $C_{70}Br_{28}$, but the structure is not known;[1] $C_{70}Br_{24}$ is predicted to be the product of highest addition level (for bulky reagents).[36]

Bromination of [60]fullerene by bromine in either CS_2 or $CHCl_3$ gives dark brown prisms of $C_{60}Br_8$.[1] The structure, determined by single crystal X-ray diffraction is shown in Fig. 7.6, and the dispositions of the bromines is shown in the Schlegel diagram in Fig. 3.11; thus $C_{60}Br_8$ (1,4,7,10,16,19,24, 36-octabromo-1,4,7,10,16,19,24,36-octahydro[60]fullerene) is a substructure of $C_{60}Br_{24}$ and may be on the pathway for the formation of it. Two molecules of bromine are occluded in the lattice for each molecule of $C_{60}Br_8$, which is also very insoluble. Although the structure of $C_{60}Br_8$ is unambiguous, calculations predict that a different arrangement of the bromines should be energetically more stable.[37]

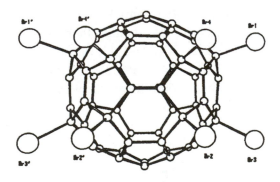

Fig. 7.6 Crystal structure for $C_{60}Br_8$.

Bromination of [60]fullerene by bromine in either CCl_4 or benzene gives magenta plates of $C_{60}Br_6$, the structure of which, determined by single crystal X-ray analysis, is shown in Fig. 7.7;[1] the Schlegel diagram is the same as for $C_{60}Cl_6$ [Fig. 7.4(a)]. The eclipsing interaction between the *cis* bromines causes the C(2)-Br bond length to be greater (2.032 Å) than the average (1.963 Å for the other C-Br bonds). There is one molecule of bromine occluded in the lattice per $C_{60}Br_6$ molecule, and it is the most soluble of the bromo[60]fullerenes.

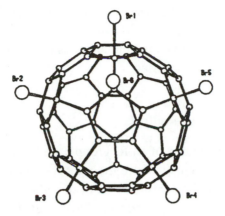

Fig. 7.7 Crystal structure for $C_{60}Br_6$.

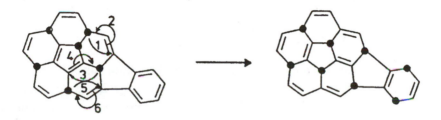

Fig. 7.8 Conjectured rearrangement of $C_{60}Br_6$ into $C_{60}Br_8$; the numbers indicate the sequence of 1,3-shifts.

All of the bromofullerenes are less stable than the chlorofullerenes, the stability order is $C_{60}Br_6 < C_{60}Br_8 < C_{60}Br_{24}$, and on heating $C_{60}Br_6$ degrades into [60]fullerene after first rearranging into $C_{60}Br_8$. This is probably triggered by the eclipsing interactions between the *cis* bromines, producing a succession of 1,3-bromine shifts, with additional bromine being acquired from the solvate, as conjectured in Fig. 7.8.

7.1.4 *Iodination*

As noted in the introduction, iodination of fullerenes has not been observed. The reason, apart from the C-I bond being the weakest of the C-halogen bonds, is almost certainly due to the size of iodine. Even 1,4-diadducts will suffer considerable steric hindrance and to avoid this the iodines would have to be further apart, thereby introducing double bonds into at least two pentagons, a process evidently too destabilising.

7.2 Reaction with Other Radicals

Reactions involving addition of an $R^\bullet$ group have been studied so far largely in the context of determining where the radical is located on the cage, how many groups add, and the conformation of the added R. The reactions have not been studied with a view to making derivatives, though this aspect may change with the improved HPLC separation techniques for fullerene derivatives now becoming available.

7.2.1 *Radical Reactions of [60]fullerene*

The simplest fullerene radical is $HC_{60}{}^\bullet$ obtained by photolysis of [60]fullerene in *t*-butylbenzene in the presence of either 2,4-dihydroxy-2,4-dimethylpentan-3-one, di-*t*-butylperoxide/propan-2-ol, or acetone/propan-2-ol. The intensity of the ESR signal depends upon the other radicals involved, which indicates that $HC_{60}{}^\bullet$ is formed here by reduction of [60]fullerene to the radical anion $C_{60}{}^{\bullet-}$ followed by protonation.[38]

Irradiation of a solution of [60]fullerene and *t*-butyl peroxide in toluene produces $C_{60}(benzyl)_n$ $n = 1 - ca.$ 15,[39]and $C_{60}Me_{34}$ if benzene is used as the solvent; with this solvent and dibenzoyl peroxide as radical initiator up to 15 phenyl groups add to the cage.[40] These results suggest that the size of the added radical governs the number that can be added. The radical adducts $C_{60}(benzyl)_n$, $n = 3$ or 5, produced after prolonged irradiation, are stable above 50°C and have been identified as the allylic $R_3C_{60}{}^\bullet$ and cyclopentadienyl $R_5C_{60}{}^\bullet$ radicals (Fig. 7.9) and in the latter the structural motif observed in halogenation, morpholine addition, and predicted to be energetically favourable in hydrogenation [see Sec. 4.3(c)], is again observed. The unpaired electrons are highly localised, and this is because delocalisation would introduce double bonds into pentagons which is unfavourable. The stability of these radicals was attributed to steric protection of the unpaired electron,[40] but this is unlikely

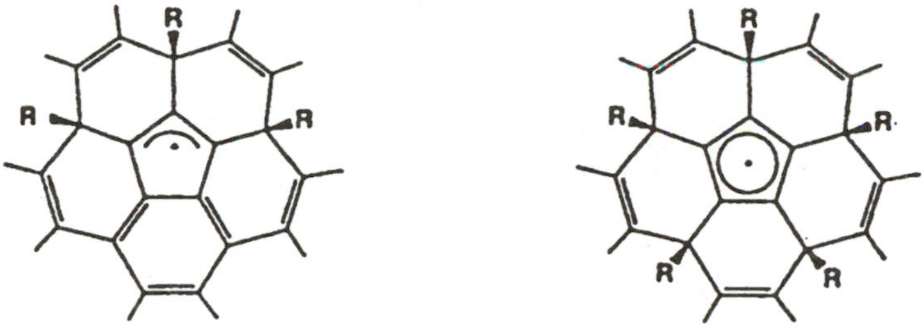

Fig. 7.9 Allylic and cyclopentadienyl radicals formed from [60]fullerene (R = benzyl).

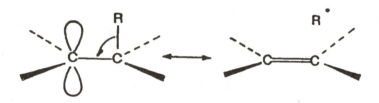

Fig. 7.10 Stabilisation of radicals in fullerenes due to sterically enforced C-C hyperconjugation.

to be correct because the corresponding species obtained in bromination (where the steric protection would be greater) are unstable and must add a further bromine to overcome this. The stability of the alkyl radicals undoubtedly arises from steric facilitation of C-C hyperconjugation,[41] which in fullerenes must be greater than in any other molecule because the electrons of the C-benzyl σ-bond and the orbital containing the unpaired electron are *forced* to be parallel (Fig. 7.10), the same will be true of C-H hyperconjugation where this can apply.

t-BuC$_{60}$• has been generated by irradiating [60]fullerene solutions with either t-butyl bromide, pivaldehyde, or di-t-butyl ketone, by reaction of [60]fullerene with di-t-butylmercury, and by oxidation of C$_{60}$Ht-Bu with iodine in THF[42,43]). The ESR spectrum indicates delocalisation of the radical primarily to the 2-, 4- and 6-positions (Fig. 7.11), the spin densities being estimated as *ca.* 0.33, 0.17 and 0.17, respectively; the higher former value reflects the fact that only in this case is a double bond not introduced into a pentagon. The signal intensity *increases* with increasing temperature, the opposite to expectation.[43] This has been attributed to the formation of dimers which

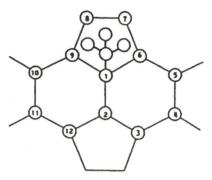

Fig. 7.11 Delocalisation sites for the radicals $RC_{60}^{\bullet}$.

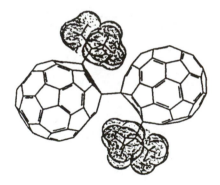

Fig. 7.12 Delocalisation of $RC_{60}^{\bullet}$ radicals at the 4-positions.

dissociate on raising the temperature. This is a general phenomenon giving structures such as shown in Fig. 7.12, and the estimated bond strengths of the dimers depends somewhat upon the sizes of the groups R, the enthalpies of formation (kcal mol^{-1}) being: (R =) *i*-Pr, 35.5; *t*-Bu, 22.0; adamantyl, 21.6; CCl$_3$, 17.1; CBr$_3$, 17.0 Steric considerations dictate that dimerisation must involve bonding between the 4-positions on each of the two cages involved.

The eclipsing interactions between the C-Me bonds of the *t*-butyl group and the C1-C2, C1-C6 and C1-C9 bonds causes free rotation of the *t*-butyl group to be hindered at low temperature.[43] Similarly, in *i*-PrC$_{60}^{\bullet}$ and EtC$_{60}^{\bullet}$ the preferred conformers locate the methyls and hydrogens respectively, over the centres of the hexagons adjacent to the point of attachment.[44] All of the fluoro analogues, $F_3CC_{60}{}^\bullet$, $F_3CF_2CC_{60}{}^\bullet$, $(F_3C)_2FCC_{60}{}^\bullet$ and $(F_3C)_3CC_{60}{}^\bullet$, adopt symmetrical conformations (with barriers to rotation of *ca.* 3.5 –7 7 kcal mol^{-1}) except the perfluoro-*i*-propyl radical which has a skewed conformation, the lone fluorine lying slightly off-centre over one of the hexagons;[45] Since F is less electron-withdrawing than CF_3 it could be expected to lie over the more electron-deficient pentagon. This reversal of expectation has been confirmed with studies of partially fluorinated methyl-, ethyl-, and *i*-propyl radicals[46] which show that the preference for location over the pentagon follows the order $CF_3 > F > H > CH_3$, which is the exact opposite of what would be expected from their inductive σ-values; other unidentified factors must operate.

Perfluoroalkylfullerenes are the most volatile fullerene derivatives known, can be sublimed readily, are stable up to at least 270°C, and are stable towards both sodium hydroxide and sulphuric acid. They have been prepared in various ways which produce a variable number of groups attached to the cage, but with an upper limit of *ca.* 16, but none of the structures are yet known. For example, reaction of solutions of perfluoropropionyl peroxide $C_2F_5C(O)OO(O)C_2F_5$ in Freon 113 with [60]fullerene at -25°C results in the attachment of 9–16 perfluoroethyl groups.[47] Heating a 1,2,4-trichlorobenzene solution of perfluoroalkyl iodide with [60]fullerene at 200°C results in addition of mainly 10 perfluorohexyl groups, though with some hydrogen incorporation as a result of abstraction from the solvent.[47] On the other hand irradiation of a benzene solution of [60]fullerene with trifluoromethyl iodide produces the addition of up to 13 trifluoromethyl groups, though with much greater hydrogen incorporation here.[47] Reaction of [60]- and [70]fullerenes with trifluoromethyl iodide and copper produces a hexane-soluble, and a hexane-insoluble fraction from each. The soluble fractions comprise derivatives with 16- and 18-trifluoromethyl groups attached to the respective cages, whilst the insoluble fractions contain in addition, a number of hydroxy groups.[48] Reaction of bis(trifluoromethyl)nitroxide radicals with [60]fullerene gives a product $C_{60}[CF_3)_2NO]_n$ where $n = 18$ on average.[49]

1-R-1,2-Dihydro[60]fullerenes, where R = CF_2CO_2Et, *n*-C_6F_{13}, CF_2Br, *n*-$C_{12}F_{25}$ and $(CF_2)_6I$ can be prepared by heating benzene solutions of [60]fullerene with Bu_3SnH and fluoroalkyl halides under reflux.[50]

Apart from $Cl_3CC_{60}{}^\bullet$, the only other chloroalkylfullerene radical examined is $(Cl_3C)_2ClC_{60}{}^\bullet$, which is stable in the dark;[51] in general it could be expected that the larger the halogen, the less stable will be the derivatives.

Radicals formed between [60]fullerene and elements other than carbon can also be made. For example, $MeSC_{60}{}^\bullet$ and $EtSC_{60}{}^\bullet$ are formed by irradiating benzene solutions of [60]fullerene with MeSSMe and EtSSEt, respectively. These behave differently from the carbon radicals described above in that the ESR signal intensity *decreases* with increasing temperature and this may arise from the weaker fullerene-S bond.[52] Use of diphenyl disulphide as reagent produced species $C_{60}Ph_x(SPh)_yH_z$ suggesting that here both $PhS^\bullet$ and $Ph^\bullet$ radicals attach to the cage.[52] Alkoxy radicals, generated from either di-*t*-butyl peroxide, di-cumyl peroxide, and bis(trifluoromethyl) peroxide also attach to the cage giving, respectively, $t\text{-}BuOC_{60}{}^\bullet$, $Me_2PhCOC_{60}{}^\bullet$ and $CF_3OC_{60}{}^\bullet$ derivatives.[52] Direct generation of alkoxy radicals from peroxide precursors is however limited by the hazards of using these latter. These difficulties can be overcome by using dialkoxy disulphide precursors which allowed the formation of $ROC_{60}{}^\bullet$ where R = Me, Et, *i*-Pr and *t*-Bu. Notably, the hyperfine coupling constants between hydrogens of the 1-ethyl and 1-*i*-propyl groups and the radical at position 2 differ between the O and S series, attributable to the differences in the C-O and C-S bond lengths.[53]

Radicals with boron in the form of the $m\text{-}B_{10}H_9C_2H_2$ (*m*-carboranyl — the boron atoms are *meta* to each other) in which the carborane cage is attached to [60]fullerene via boron are known. These also exhibit dimerisation in the manner shown in Fig. 7.12, and the dimerisation rate coefficient is $1 \times 10^6\ M^{-1}\ s^{-1}$ at 27°C.[54] Irradiation of a benzene solution of trimethylaminoboron and [60]fullerene in the presence of di-t-butyl peroxide produces $Me_3H_2BC_{60}{}^\bullet$ adducts which dimerise with a rate coefficient of *ca.* $2.6 \times 10^6\ M^{-1}\ s^{-1}$ at 27°C.[55] Phosphorus-containing radicals e.g. $(RO)_2OPC_{60}{}^\bullet$ (R - Me, Et, *i*-Pr, from irradiation of R_2Hg) can be prepared; the rate of dimerisation of the *i*-propyl derivative is $1.9 \times 10^6\ M^{-1}\ s^{-1}$ at 0°C; longer reaction times result in five groups adding to the cage (*cf.* Fig. 7.9). These phosphonyl radicals have discreet conformations and the kinetic and thermodynamic parameters for the hindered rotations can be determined.[56]

Rhenium-containing radicals $(CO)_5Re^\bullet$, produced either by photodissociation of $Re(CO)_{10}$, or by reaction of $(\eta^3\text{-}Ph_3C)Re(CO)_4$ with CO (which displaces $PPh_3C^\bullet$), combine with [60]fullerene to give $C_{60}[Re(CO)_5]_2$, believed to have a 1,4-addition pattern. This is unstable and decomposes probably via $(CO)_5ReC_{60}{}^\bullet$.[57]

More complex radicals are known in which the fullerene contains other addends. Examples include the addition of either $Pd(PPh_3)_2$ or $Pt(PPh_3)_2$ to $(RO)_2OPC_{60}{}^\bullet$ and η^2-addition of $Pt(PPh_3)_4$ to $[(RO)_2OPC_{60}]_2$ giving derivatives which dissociate to the monomeric radicals $(RO)_2OP.Pt(PPh_3)_2C_{60}{}^\bullet$ when exposed to light. In these species the $Pt(PPh_3)_2$ group is located in different positions relative to the radical centre, producing five distinct ESR signals.[58] Even more complex ESR spectra, consisting of seven overlapping components, are obtained from the addition of dialkoxyphosphoryl radicals to di(4-methoxyphenyl)methano[60]fullerene.[59]

7.2.2 *Radical Reactions of [70]fullerene*

Reaction of radicals with [70]fullerene creates the possibility of interesting regioisomerism since there are five distinct carbon atoms to which the radical might attach. For example addition of hydrogen atoms (from irradiation of benzophenone/*i*-propanol in benzene) gives five different radicals, four of which derive from the addition of one hydrogen, the other from addition of three hydrogens.[60] Three different spin adducts are also obtained for $HC_{70}{}^\bullet$ generated from irradiation of a solution of [70]fullerene/benzene/cyclohexadiene.[61]

t-$BuC_{70}{}^\bullet$ consists of a mixture of three isomers having g values of 2.00271, 2.00248 and 2.00210, the corresponding hyperfine interactions with the C-atom attached to the cage being 14.2, 11.3 and 13.8 G. Likewise three isomers were observed for $Cl_3CC_{70}{}^\bullet$ (34.6, 30.5 and 26.7 G) and $(MeO)_2OPC_{70}{}^\bullet$ (71.2, 66.8 and 55.9 G).[62] For the t-butyl radical, only the isomer of g 2.0021 (13.8 G) showed hyperfine interactions with the carbons *ortho* and *para* to the radical centre comparable in magnitude to those observed for t-$BuC_{60}{}^\bullet$ (*cf.* Fig. 7.11). Since only the structure of the end cap parallels that in [60]fullerene, and moreover that this radical is the only one to show an increased ESR signal with increasing temperature (indicating dimerisation), it is probable that the t-butyl group in this radical is attached to C2 (or *a* carbon — see Fig. 2.2) with the radical centred on C1 (*b* carbon). This needs further confirmation, since the addition (across what is usually the most reactive bond in [70]fullerene), could in principle take place in the reverse direction.

Five spin adducts have been observed for $MeOC_{70}{}^\bullet$ [assumed to take place at each of the positions (*a–e*, Fig. 2.2)], three for $MeSC_{70}{}^\bullet$ and five also for

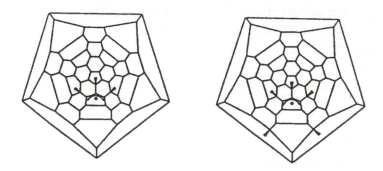

Fig. 7.13 Proposed structures of $[(MeO)_2OP]_3C_{70}{}^{\bullet}$ and $[(MeO)_2OP]_5C_{70}{}^{\bullet}$.

$F_3CC_{70}{}^{\bullet}$, but no definitive assignments are yet possible.[63] Three spin adducts have been observed for the *m*-carboranylC$_{70}{}^{\bullet}$ radical, two of which decay within 1 s of the removal of light, whereas the third one has a lifetime of more than 10 min, but again no assignments have been made.[64] Four different types of spin adducts are obtained for (*i*-PrO)$_2$OPC$_{70}{}^{\bullet}$, one of which has *g* and G factors similar to those for the [60]fullerene analogue, and is therefore assigned to addition at the *a* carbon; two others are assigned to addition at either of the *b*, *c*, or *d* carbons; whilst the fourth seems to arise either from recombination of biradicals with a monoradical, or from dimerisation of diradicals.[65] More prolonged reaction leads to the formation of allylic radicals as shown in Fig. 7.13.[66]

References

1. P. R. Birkett, P. B. Hitchcock, H. W. Kroto, R. Taylor and D. R. M. Walton, *Nature*, **357** (1992) 479.
2. H. W. Kroto, J. R. Heath, S. C. O'Brien, R. F. Curl and R. E. Smalley, *Nature*, **318** (1985) 162.
3. S. K. Chowdhury *et al.*, *Org. Mass Spectrom.*, **28** (1993) 860.
4. H. Selig, K. Kniaz, G. B. M. Vaughan, J. E. Fischer and A. B. Smith, *Macromol. Symp.*, **82** (1994) 89.
5. R. Taylor *et al.*, *Nature*, **355** (1992) 27; *J. Chem. Soc., Chem. Commun.*, (1992) 665.

6. A. A. Tuinman, A. A. Gakh, J. L. Adcock and R. N. Compton, *J. Am. Chem. Soc.*, **115** (1993) 5885.

7. O. V. Boltalina, A. K. Abdul-Sada and R. Taylor, *J. Chem. Soc., Perkin Trans. 2*, (1995) 981.

8. R. Taylor *et al.*, *J. Chem. Soc., Perkin Trans. 2*, (1995) 181.

9. J. H. Holloway *et al.*, *J. Chem. Soc., Chem. Commun.*, (1991) 966.

10. H. Selig *et al.*, *J. Am. Chem. Soc.*, **113** (1991) 5475.

11. R. Taylor and D. R. Walton, *Nature*, **363** (1993) 685.

12. H. Hamwi *et al.*, *Fullerene Sci. & Technol.*, **1** (1993) 499.

13. A. A. Tuinman, P. Mukherjee, J. L. Adcock, R. L. Hettich and R. N. Compton, *J. Phys. Chem.*, **96** (1992) 7584.

14. R. Taylor *et al.*, *J. Chem. Soc., Chem. Commun.*, (1993) 875.

15. K. Kniaz *et al.*, *J. Am. Chem. Soc.*, **115** (1993) 6060.

16. O. V. Boltalina *et al.*, *J. Chem. Soc., Perkin Trans. 2*, (1996) 2275.

17. A. A. Gakh, A. A. Tuinman, J. L. Adcock, R. A. Sachleben and R. A. Compton, *J. Am. Chem. Soc.*, **116** (1994) 819.

18. A. J. Adamson, J. H. Holloway, E. G. Hope and R. Taylor, *Fullerene Sci. & Technol.*, **5** (1997) 629.

19. O. V. Boltalina, A. Ya. Borschevskii, L. N. Sidorov, J. M. Street and R. Taylor, *Chem. Commun.*, (1996) 529.

20. O. V. Boltalina, J. M. Street and R. Taylor, *J. Chem. Soc., Perkin Trans. 2*, (1998) 649.

21. O. V. Boltalina, V. Yu. Markov, R. Taylor and M. P. Waugh, *Chem. Commun.*, (1996) 2549.

22. A. G. Avent *et al.*, *J. Chem. Soc., Perkin Trans. 2*, (1998) 1319.

23. D. M. Cox *et al.*, *J. Am. Chem. Soc.*, **116** (1994) 1115.

24. R. Taylor, *J. Chem. Soc., Perkin Trans. 2*, (1994) 2497; P. W. Fowler, D. Mitchell, R. Taylor and G. Siefert, *J. Chem. Soc., Perkin Trans. 2*, (1997) 1901.

25. O. V. Boltalina, J. M. Street and R. Taylor, unpublished work.

26. R. Hettich, C. Jin and R. N. Compton, *Int. J. Mass Spectrom. Ion Proc.*, **138** (1994) 263.

27. O. V. Boltalina, L. V. Sidorov, E. V. Sukhanova and I. D. Sorokin, *Chem. Phys. Lett.*, **230** (1994) 567.

28. C. Jin *et al.*, *Phys. Rev. Lett.*, **73** (1994) 1882.

29. A. A. Gakh, A. A. Tuinman, J. L. Adcock and R. N. Compton, *Tetrahedron Lett.*, **34** (1993) 7167.
30. R. Taylor, unpublished work.
31. R. Taylor, *Electrophilic Aromatic Substitution*, Wiley, Chichester, 1989, pp. 363–364.
32. G. A. Olah *et al.*, *J. Am. Chem. Soc.*, **113** (1991) 9385.
33. F. N. Tebbe *et al.*, *J. Am. Chem. Soc.*, **113** (1991) 9900; *Science*, **256** (1992) 822.
34. P. R. Birkett, A. G. Avent, A. D. Darwish, H. W. Kroto, R. Taylor and D. R. M. Walton, *J. Chem. Soc., Chem. Commun.*, (1993) 1230.
35. P. R. Birkett, A. G. Avent, A. D. Darwish, H. W. Kroto, R. Taylor and D. R. M. Walton, *J. Chem. Soc., Chem. Commun.*, (1995) 683.
36. R. Taylor, *J. Chem. Soc., Perkin Trans. 2*, (1993) 813.
37. P. W. Fowler and J. P. B. Sandall, *J. Chem. Soc., Perkin Trans. 2*, (1995) 1247.
38. C. N. McEwen, R. G. McKay and B. S. Larsen, *J. Am. Chem. Soc.*, **114** (1992) 4412.
39. P. J. Krusic, E. Wasserman, P. N. Keizer, J. R. Morton and K. F. Preston, *Science*, **254** (1991) 1183.
40. P. J. Krusic *et al.*, *J. Am. Chem. Soc.*, **113** (1991) 6274.
41. R. Taylor, *Electrophilic Aromatic Substitution*, Wiley, Chichester (1989), pp. 15–18.
42. P. J. Fagan, P. J. Krusic, D. H. Evans, S. A. Lerke and E. A. Johnston, *J. Am. Chem. Soc.*, **114** (1992) 9697.
43. J. Morton, K. F. Preston, P. J. Krusic, S. A. Hill and E. Wasserman, *J. Am. Chem. Soc.*, **114** (1992) 5454; *J. Phys. Chem.*, **96** (1992) 3576.
44. P. J. Krusic, D. C. Roe, E. Johnston, J. R. Morton and K. F. Preston, *J. Phys. Chem.*, **97** (1993) 1736.
45. J. R. Morton and K. F. Preston, *J. Phys. Chem.*, **98** (1994) 4993; J. R. Morton, F. Negri, and K. F. Preston, *Chem. Phys. Lett.*, **232** (1995) 16.
46. J. R. Morton, F. Negri, K. F. Preston and G. Ruel, *J. Phys. Chem.*, **99** (1995) 10114.
47. P. J. Fagan, P. J. Krusic, C. N. McEwen, J. Lazar, D. H. Parker, N. Herron, and E. Wasserman, *Science*, **262** (1993) 404.
48. J. D. Crane, H. W. Kroto, G. J. Langley and R. Taylor, unpublished work.

49. D. Brizzolara, J. T. Ahlemann, H. W. Roesky and K. Keller, *Bull. Soc. Chim. Fr.*, **130** (1993) 745.

50. M. Yoshida, D. Suzuki and M. Iyoda, *Chem. Lett.*, (1996) 1097.

51. J. R. Morton, F. Negri and K. F. Preston, *Canad. J. Chem.*, **72** (1994) 776.

52. M. A. Cremoni, L. Lunazzi, G. Placucci and P. J. Krusic, *J. Org. Chem.*, **58** (1993) 4735.

53. R. Borghi, L. Lunazzi, G. Placucci, G. Cerioni and A. Plumitallo, *J. Org. Chem.*, **61** (1996) 3327.

54. B. L. Tumanskii, V. V. Bashilov, S. P. Solodnikov, N. N. Bubnov and V. I. Sokolov, *Russ. Chem. Bull.*, **44** (1995) 1771.

55. B. L. Tumanskii *et al.*, *Russ. Chem. Bull.*, **43** (1994) 624.

56. B. L. Tumanskii, V. V. Bashilov, N. N. Bubnov, S. P. Soldnikov and A. A. Khodak, *Russ. Chem. Bull.*, **43** (1994) 1582.

57. S. Zhang, T. L. Brown, Y. Du and J. R. Shapley, *J. Am. Chem. Soc.*, **115** (1993) 6705.

58. B. L. Tumanskii, V. V. Bashilov, V. V. Bubnov, S. P. Solodnikov and V. I. Sokolov, *Russ. Chem. Bull.*, **43** (1994) 884.

59. B. L. Tumanskii, M. N. Nefedov, V. V. Bashilov, S. P. Solodnikov, N. N. Bubnov and V. I. Sokolov, *Russ. Chem. Bull.*, **45** (1996) 2865.

60. J. R. Morton, F. Negri and K. F. Preston, *Chem. Phys. Lett.*, **218** (1994) 467.

61. R. Borghi, L. Lunazzi, G. Placucci, P. J. Krusic, D. A. Dixon and L. B. Knight, *J. Chem. Phys.*, **21** (1994) 5395.

62. P. N. Keizer, J. R. Morton and K. F. Preston, *J. Chem. Soc., Chem. Commun.*, (1992) 1259.

63. R. Borghi, B. Guidi, L. Lunazzi and G. Placucci, *J. Org. Chem.*, **61** (1996) 5667.

64. B. L. Tumanskii, V. V. Bashilov, S. P. Solodnikov, N. N. Bubnov and V. I. Sokolov, *Russ. Chem. Bull.*, **44** (1995) 1771.

65. B. L. Tumanskii, V. V. Bashilov, N. N. Bubnov, S. P. Solodnikov and V. I. Sokolov, *Bull.Acad. Sci., USSR*, **41** (1992) 1521.

66. B. L. Tumanskii, V. V. Bashilov, N. N. Bubnov, S. P. Solodnikov and V. I. Sokolov, *Russ. Chem. Bull.*, **42** (1993) 203.

8

Nucleophilic Substitution of Fullerenes: Fullerenes as Electrophiles

Nucleophilic substitution in fullerenes occurs very readily, but for making specific derivatives has been less utilised than some other reactions. This is because most studies involve replacement of e.g. halogens that have been added to the cage via radical reactions, which are difficult to control (Chap. 7) (it is not possible to add halogens to the cages under electrophilic conditions). The majority of substitutions that have been studied involve either formation of hydroxyfullerenes (fullerenols) or replacement of halogen by aryl groups, which is also an electrophilic aromatic substitution. Fullerenes are by far the strongest carbon electrophiles known, and even if unhalogenated will substitute aromatics electrophilically; the reaction is faster if halogen on the fullerene is being replaced; for completeness, the direct substitution of fullerenes into aromatics (*fullerenylation*) is included at the end of this chapter. The ease of halogen replacement follows the order F > Cl > Br which parallels the reactivity of halogenoaromatics towards nucleophilic aromatic substitution, and is also the order of halogenoalkanes as electrophiles in electrophilic aromatic substitution.[1] There have been no reported studies of replacement of bromine in bromofullerenes because they are very insoluble, but these can be produced *in situ* and then substituted without isolation.

The mechanism of these substitutions are different from any previously encountered in organic chemistry. A normal S_N2 reaction is not possible because it would involve backside attack, i.e. from *within* the cage. An S_N1 reaction seemed to be highly improbable because of the difficulty of forming a cation from the parent fullerene cages. For this reason an addition-elimination mechanism was proposed for fluorine replacement,[2] and this requires the

114

incoming group to occupy a different position on the cage from that of the departing group. More recent work has shown that for the Lewis-acid catalysed replacement of chlorine by phenyl (Sec. 8.6.1), the phenyl group occupies the *same* position as that of the departing chlorine, so nucleophilic substitution involving frontside attack on a preformed cation (or partial cation) must occur in cases where the loss of halogen is aided by the catalyst. Moreover, in the presence of $AlCl_3$, the intermediate carbocation can be isolated and characterised spectroscopically.[3] It should be noted that it is easier to form a cation when there are addends on the cage because the electron withdrawal is reduced by the presence of sp^3-hybridised carbons (which are less electronegative than sp^2 carbons).

8.1 Methoxydehalogenation

The reaction of polychloro[60]- or [70]fullerenes with methanol/KOH gives a product showing a broad envelope of methoxy groups at δ *ca.* 3.7 in the ^1H NMR spectrum, and FAB mass spectrometry indicates the presence of *ca.* 26 methoxy groups.[4] A similar NMR pattern is observed in the products of reaction of both fluorofullerenes[5] and $C_{60}Cl_6$ with methoxide ion/methanol,[6] but in none of these cases have specific derivatives yet been isolated.

8.2 Fullerenol Formation

8.2.1 *Hydroxydenitration and Hydroxydesulphonation*

Three procedures involving nucleophilic substitution have been utilised for the formation of fullerenols, an important class of *water-soluble* fullerene derivatives which are precursors for preparation of polymers. Characterisation of fullerenols presents exceptional difficulties, and the mechanism of their formation is still open to some uncertainties, but nucleophilic substitution is evidently involved during the process. Fullerenols are most unusual compounds in that the aqueous solubility is *increased* in acid and *decreased* in base, the opposite of what would be expected. Reasons for this are suggested below.

The methods of formation each involve initial addition of a group to the cage under conditions normally associated with electrophilic aromatic substitution (but which here may involve single electron transfer processes in view of the reluctance of the unaddended cages to form cationic intermediates). The conditions involve initial reaction of the fullerene with either: (i) nitric acid/sulphuric acid (the standard 'mixed acid' conditions[7] for nitration),[8] (ii) nitronium tetrafluoroborate,[8] or (iii) fuming sulphuric acid.[9]

8.2.1.1   *Use of Aqueous* HNO_3/H_2SO_4

The optimum conditions involve aq. sulphuric acid (containing potassium nitrate to generate the nitric acid) and a temperature of 85–95°C. The difficulty with this technique is that it is difficult to eliminate the inorganic by-products because of the water-solubility of the fullerenols. However, raising the pH to > 9.0 causes them to precipitate and reasonable purification can be effected, though the presence of traces of sodium sulphate are troublesome. The products from hydroxylation of [60]fullerene by this method contain 14–15 OH groups per molecule, and are formed in *ca.* 95% yield. The mechanism has been suggested as involving intermediate unstable cationic intermediates which then undergo nucleophilic substitution by the *in situ* water to give the fullerenol. However, electron-deficient double bonds have been shown to undergo ready epoxidation by conc. nitric acid to give epoxides (and pinacones after rearrangement),[10] so given the high electron deficiency of these bonds in fullerenes, an oxidation process cannot be entirely precluded at present. It is relevant therefore that reaction of [60]fullerene with fuming nitric acid gives hygroscopic polyhydroxy-polynitro derivatives which are soluble in acid, base, and aqueous media, and have a total of *ca.* 24 groups attached to the cage.[11]

Fullerenols have a very characteristic IR spectrum (Fig. 8.1) consisting of five broad bands at around 3430, 1595, 1385, 1085 and 450–550 cm^{-1}. The X-ray photoelectron spectrum of the C_{1s} binding energy of fullerenols indicate that a proportion of the carbons are in a di-oxygenated state involving either carbonyl (but ruled out by the absence of $C = O$ in the IR), ketal (but improbable given the fullerene structure) or hemiketal, **8.1**. Support for the hemiketal structure was provided by a peak in the ^{13}C NMR spectrum at 170.3 ppm.[12] This structure potentially introduces many double bonds into pentagons, though this may be less important with many double bonds having undergone addition, thereby relieving strain. The existence of the OH groups (difficult to establish

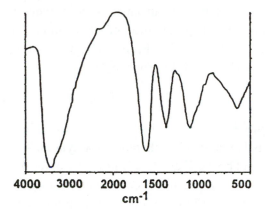

Fig. 8.1 Characteristic fullerenol IR spectrum.

8.1

by mass spectrometry because of the ease of fragmentation) has been confirmed by ester formation (IR band at 1720–1740 cm^{-1}) the product being soluble in organic solvents rather than water. Esterification is incomplete though, presumably due to steric hindrance.

MALDI mass spectrometry indicates that though the average OH content of the fullerenols produced by this method is around 15, as many as *ca.* 24 OH groups are present in some molecules.

8.2.1.2 *Use of Nitronium Tetrafluoroborate*

In this procedure a mixture of the nitronium salt and either an aryl- or alkylcarboxylic acid (which together form the acyl nitrate intermediate

$RCONO_2$, Eq. 8.1) are reacted in an organic solvent with the fullerene which is presumed to undergo addition of NO_2 and OCOR [where R = CF_3, $(CH_2)_nCH_3$, $4X$-C_6H_4 (X = H, F, Br, Cl, CN, NO_2, OMe, NMe_2), 3,5-dimethoxyphenyl and 3,5-dinitrophenyl].

$$NO_2BF_4 + RCO_2H \rightarrow RCO_2NO_2 + HBF_4 \qquad (8.1)$$

Nucleophilic replacement of nitro by hydroxyl during the work-up procedure gives ultimately polyhydroxy[60]fullerene carboxylates, the IR spectra of which show the presence of both OH and C = O groups. For the 4-bromophenyl ester, elemental analysis of the product is consistent with an average composition of 5 ester groups and 13–15 OH groups per fullerenol. Subsequent hydrolysis of the ester by aqueous alkali produces fullerenols showing the characteristic bands in the IR spectrum, and having an average composition of 18–20 OH groups per fullerenol. Given that acetic anhydride and nitric acid readily form acetyl nitrate, $RCONO_2$, which is well known to add to both aromatics and alkenes to give 1,4- and 1,2-nitroacetates,[13] use of this reagent combination ought also to lead to fullerenols, but this possibility has yet to be explored. The mass spectra (FAB/LSIMS) confirmed the presence of approximately the above number of OH groups, but is particularly notable in showing formation of higher fullerenes of regular 24 amu increments. This is a feature of this mass spectrometry technique, and has been observed by others using different compounds.[14]

8.2.1.3 *Use of Oleum*

In this method,[9] the fullerene is reacted with fuming sulphuric acid (28% SO_3) at 55–60°C under N_2. Subsequent reaction either with water at 85–90°C or aq. NaOH at room temperature yields the fullerenols, containing an average of 10–12 OH groups per fullerenol molecule This conclusion is supported by thermal gravimetry of the presumed polycyclosulphated precursors which is consistent with six such groups being attached to each fullerene cage. X-ray photoelectron spectrum of the C_{1s} binding energy of these fullerenols indicates that no carbons are in a di-oxygenated state, indicating an absence here of hemiketal structures (^{13}C NMR studies support this). However, the IR spectra are not significantly different from those obtained under the previous two

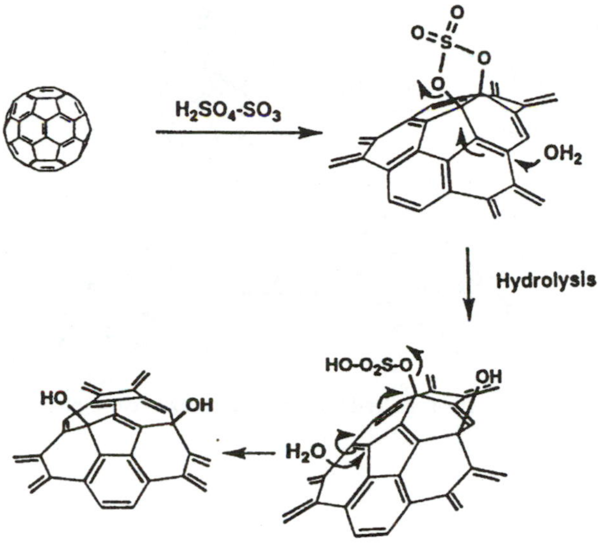

Fig. 8.2 Scheme proposed for formation of fullerenols from [60]fullerene and HSO_4/SO_3.

conditions, though the general broadness of the IR bands for fullerenols makes identification of minor differences difficult. The hydroxy groups in the product can be converted to the corresponding hydrogen sulphates by reaction with H_2SO_4/SO_3.

The proposed mechanism involves formation of a cyclosulphated intermediate which is followed by two nucleophilic replacement steps yielding the final dihydroxy derivative (Fig. 8.2) though it is probable that one of the OH groups in the final product shown would undergo both 1,3- and 1,5-shifts to give a 1,4-dihydroxy derivative, in order to minimise the number of double bonds in pentagons. An alternative viable mechanism involves sultone formation (Fig. 8.3), the argument against this being that hydrolysis would give a hydroxyfullerenesulphonic acid, which would thus still contain sulphur (not however found in the fullerenol product). Moreover, the necessary cleavage of the C-S bond in sulphonic acids and derivatives is very slow, so the conversion of the intermediate hydroxyfullerenesulphonic acid to fullerenol should likewise be slow in contrast to the rapid product formation. Nevertheless, polysultones [of average structure $C_{60}(SO_3)_{5.2}$] have been shown to be formed from the reaction between [60]fullerene and either neat SO_3 or fuming sulphuric acid.[15]

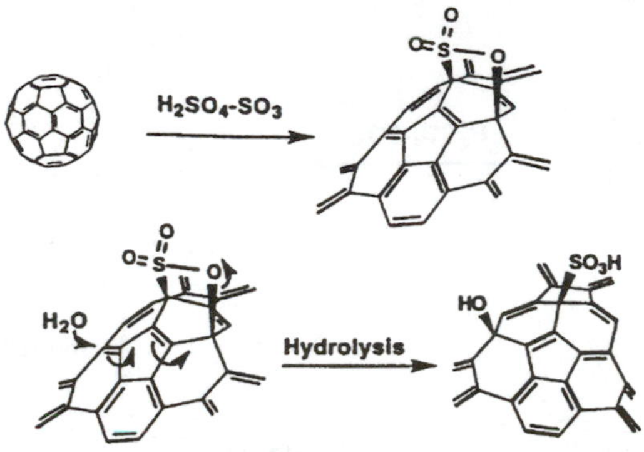

Fig. 8.3 Mechanistic pathways that can lead to either sultones or cyclosulphates, from the reaction between fullerenes and SO_3.

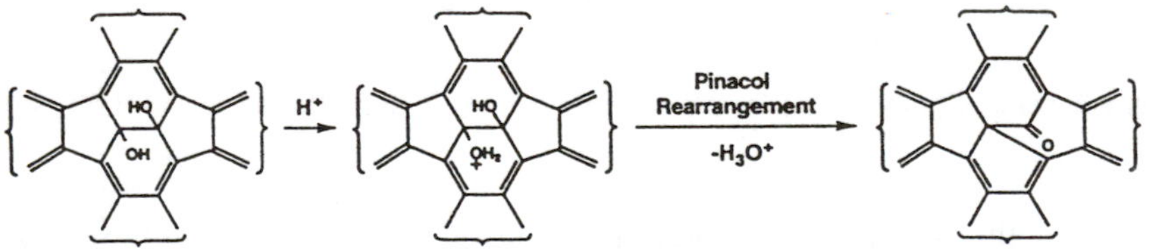

Fig. 8.4 Mechanism for acid-catalysed irreversible rearrangement of a 1,2-diol, giving a pinacolone.

The first step in the overall reaction evidently involves one-electron oxidation of the fullerene by SO_3 giving a radical cation (indicated by the immediate formation of a green EPR-active solution), which is then slowly converted to a cation, accompanied by formation of an orange precipitate and loss of EPR activity.

Further investigation into the hemiketal structure involves reacting the fullerenols, produced by the nitric acid/sulphuric acid method, with dilute acid, and then with base. This distinguishes between a 1,2-diol (which acid converts irreversibly to a pinacolone, Fig. 8.4), and a hemiketal, which produces a hydroxyketone. However, the latter, after treatment with base should convert back to the hemiketal Fig. 8.5), which IR studies confirm to be the case; moreover the intermediate hydroxyketone shows a band at 1722 cm^{-1} in the IR.

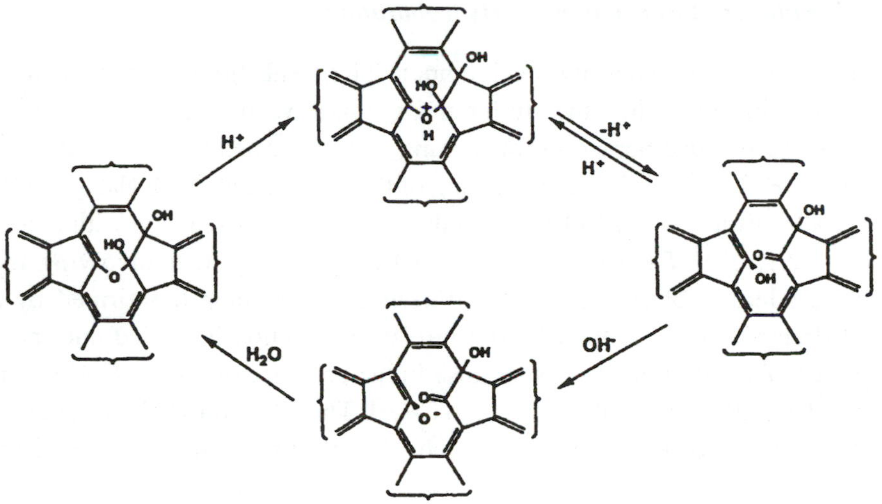

Fig. 8.5 Mechanism for acid- and base-catalysed interconversion of a hemiketal and a hydroxyketone.

8.2.2 *Fullerenols from Nitrofullerenes*

Reaction of [60]fullerene in toluene with N_2O_4 ($NO_2^\bullet$) followed by precipitation with hexane (presumably wet) gives nitrofullerenols having an average composition of 6–8 nitro groups and 7–12 hydroxy groups per [60]fullerene cage.[16] The polynitro intermediates (which are orange-brown and also readily formed using $HNO_3/NaNO_2$ to generate $NO_2^\bullet$) are soluble in a range of common organic solvents. Subsequent nucleophilic substitution of these intermediates by reaction with aq. NaOH produces fullerenols giving the characteristic IR spectrum, and having ≥ 16 OH groups per [60]fullerene cage.[17]

8.2.3 *Fullerenol Formation from Halogenofullerenes*

Mono- and dihydroxyfullerenes are obtained in nucleophilic substitution of halogenofullerenes, especially chloro- and fluorofullerenes, by water. This is described in more detail in Secs. 8.6.1–8.6.4.

8.2.4 *Fullerenol Formation via Hydroboration*

Whilst this process does not involve nucleophilic substitution, it is included here together with the other fullerenol preparative methods. Reaction of [60]fullerene with diborane and quenching of the intermediate with acetic acid gives $C_{60}H_2$ (Sec. 4.1.5), but quenching of the intermediate with H_2O_2/NaOH gives a [60]fullerenol [IR identical to that obtained from reaction with oleum (Sec. 8.2.1.3)].[18] The expected product here would be that involving the normal addition of H and OH, but this is only obtained (confirmed by a C-H stretching band in the IR spectrum) if the reaction is carried out under nitrogen. The difference arises because in the presence of air, the C-H bonds formed initially are rapidly oxidised to C-OH. This rapid (allylic) oxidation of fullerene C-H bonds is now a well established feature of fullerene chemistry (see Chap. 4).

The product shows the normal fullerenol properties of solubility only in water (slightly), HCl, Me_2SO and pyridine, and can be esterified giving a product soluble in organic solvents, but not aqueous solutions.[18]

8.2.5 *Fullerenol Formation Involving Reaction with Oxygen*

Fullerenes can be hydroxylated by $NaOH/O_2$ in a two-phase (toluene/H_2O) solution, using n-Bu_4NOH as a phase-transfer catalyst.[19]

8.3 Biological Applications of Fullerenols

[60]Fullerenols show excellent efficiency (up to 80%) in scavenging superoxide radical species $O_2^{\bullet-}$ produced in the model experiments from xanthine and xanthine oxidase,[20] and are more effective than the spin-trapping reagent, 5,5′-dimethyl-1-pyrrolidine-N-oxide (DMPO).[21] These radicals are thought to play an important role in many biological malfunctions, including pathological processes, various cytoxic injuries, ageing, carcinogenesis etc. [60]Fullerenols have been shown to reduce the concentrations of these radicals in blood of patients having gastric cancer,[22] and have an antiproliferative effect on vascular smooth muscle cells,[23] which are major factors in the development of antherosclerosis and restenosis. They have also been shown to have

DNA-cleaving ability, and to exhibit cytoxicity against tumor cells, the effect upon the latter being enhanced 25-fold by light.[24]

8.4 Polymers Derived from Fullerenols

These are described in Chap. 12.

8.5 Allyldechlorination

Reaction of $C_{60}Cl_6$ (see Sec. 7.1.2) with allyl bromide and ferric chloride, results in replacement of all six chlorines by allyl groups. By contrast, in the arylation reactions described below only *five* of the chlorines are replaced, the most inaccessible one being unaffected. The difference arises because the allyl group is less sterically demanding than a phenyl ring. This also provides indirect evidence that *ipso* substitution of the chlorines occurs (rather than an addition elsewhere followed by elimination) and NMR analyses of the product has provided direct confirmation of the replacement mechanism.[25]

8.6 Aryldehalogenation

This constitutes one of the largest classes of fullerene reactions: in the presence of a Lewis acid catalyst, the fullerene cages carries out electrophilic substitution into aromatics. In view of the number of reaction sites on the fullerenes and the number of aromatics and sites for substitution in them, the number of derivatives is vast. So far reactions have been limited to substitution of chlorines in $C_{60}Cl_6$ and $C_{70}Cl_{10}$ (see Sec. 7.1.2), of fluorine in $C_{60}F_{18}$ (see Sec. 7.1.1.2) and to substitution of bromine in brominated [60]fullerene (produced *in situ* from bromine and [60]fullerene). These reactions thus constitute a typical aromatic alkylation using an alkyl halide. Likewise just as alkylation of aromatics can be performed using an alkene and a Lewis acid catalyst, so the corresponding reactions using the parent fullerene as alkene have also been studied and are included here for comparison purposes.

The reactions show all the typical characteristics of an electrophilic aromatic substitution. Thus the effectiveness of catalysts is $AlCl_3 > FeCl_3$, the reaction is faster with more reactive aromatics, e.g. toluene is more reactive than benzene, and the reaction is subject to steric hindrance which is here severe because of the bulk of the fullerene electrophile. Hence mesitylene will not react because there is no site sufficiently unhindered in the molecule. In these reactions, [60]fullerene is more reactive than [70]fullerene, and this is a pattern noted in many other reactions such as with benzyne and cyclopentadiene (Chap. 9) and hydrogenation (Chap. 4).

8.6.1 *Substitutions in $C_{60}Cl_6$*

If $C_{60}Cl_6$ is reacted with benzene and ferric chloride then $C_{60}Ph_5Cl$ (Fig. 8.6) is obtained. Further reaction of this derivative with PPh_3 results in replacement of the remaining chlorine by hydrogen, derived from traces of water in the solvent.[26]

The reaction has been extended[27] to include reaction with anisole, t-butylbenzene and fluorobenzene (which substitute in the para positions for steric and electronic reasons), toluene (which substitutes both *ortho* and *para*) and thiophene (which appears to react at both 2- and 3-positions). An interesting illustration of the importance of steric hindrance in these reactions is provided

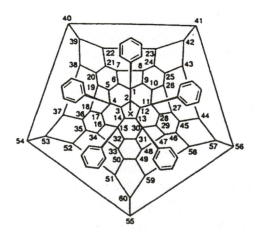

Fig. 8.6 Schlegel diagram of the product (X = Cl) of racting $C_{60}Cl_6$ with benzene/FeCl₃; reaction of this with PPh_3 gives the hydride (X = H).

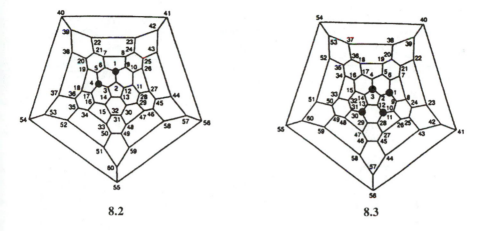

8.2   8.3

by the reaction with trimethylsilylbenzene, which would normally involve C-Si cleavage rather than C-H cleavage.[28] However, in the reaction with the fullerene the latter occurs only, due to the difficulty of the fullerene cage approaching the aromatic *ipso* ring carbon adjacent to the bulky trimethylsilyl group. The strong electron withdrawal by the fullerene also makes it difficult to cleave the C-Si bond in the product, whereas such cleavage is normally a facile electrophilic aromatic substitution.

Various other products are obtained in this reaction. These include symmetrical $C_{60}Ph_2$, **8.2**, and an unsymmetrical $C_{60}Ph_4$ derivative, the structure of which is believed to be **8.3**.[29] Reaction of $C_{60}Ph_5Cl$ with $AlCl_3$ followed by traces of water produces two fullerenols, one of which, **8.4**, is symmetric and results from a 1,2-shift of a phenyl group. This confirms that a carbocation is formed on the cage in these substitutions, resulting in the potential creation of an anti-aromatic (4π-electron) five-membered ring. To avoid this, the adjacent phenyl group carries out a 1,2-shift, synchronous with the departure of the chlorine, and gives a much more stable carbocation (recently isolated and characterised).[3] An OH group then attaches at the carbocationic centre and hence occupies the position vacated by the phenyl group giving **8.4** (Fig. 8.7).[3] The NMR spectra indicate that the unsymmetric fullerenol has the structure **8.5**, believed to result from a 1,3-shift of the intermediate carbocation, followed by attack of the OH group at the new carbocationic centre, and then a 1,5-shift of the phenyl group in the central pentagon. The rearrangement (Fig. 8.7) confers two advantages: the Ph and OH groups then become adjacent across a

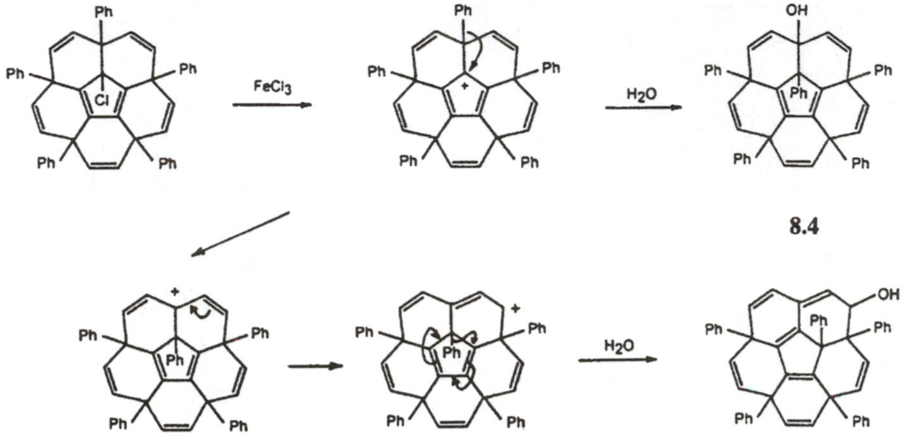

Fig. 8.7 Conjectured mechanism leading to formation of symmetrical (**8.4**) and unsymmetrical (**8.5**) $C_{60}Ph_5OH$.

longer 6:5 bond rather than a shorter 6:6 bond, and the phenyl migration creates a more stable *p*-quinonoid structure.

On standing in air and light, $C_{60}Ph_5H$ forms two oxidised products. The minor component is unsymmetrical $C_{60}Ph_4O_2$ which eliminates two molecules of CO during EI mass spectrometry to give phenylated C_{58} derivatives. This observation has important ramifications concerning the way in which structure affects cage shrinkage:

(1) Loss of C_2 from a 6:5-single bond is unfavourable because the resultant dangling bonds can only be satisfied by the formation of a 4-membered ring, and two *planar* sp^3-hybridised carbons (Fig. 8.8); clearly this is impossible.

(2) Loss of C_2 from a 6:6-double bond is also unfavourable because this leads (Fig. 8.9) to the formation of two 4-membered rings (flanking an 8-membered ring), and the resultant cage will be considerably strained.

(3) Loss of C_2 from a 6:5-*double* bond is much less unfavourable because this produces a 7-membered ring flanked by four 5-membered ones (Fig. 8.10). In an unaddended fullerene, double bonds are not normally located

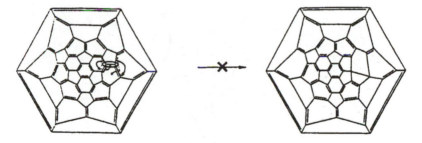

Fig. 8.8 The structure that would result form C$_2$ elimination across a 6:5-single bond.

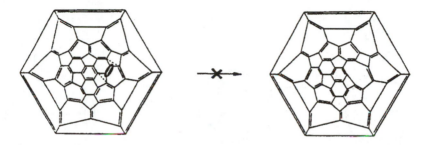

Fig. 8.9 The structure that would result from C$_2$ elimination across a 6:6-double bond.

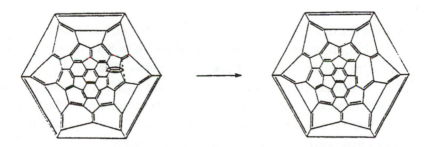

Fig. 8.10 The structure that results from C$_2$ elimination across a 6:5-*double* bond.

across such a bond. However, in cases where this arises due to the presence of addends, and where these are located in the created five-membered rings, thereby reducing the strain, *cage shrinkage can be expected to occur on heating.*

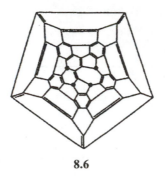

8.6

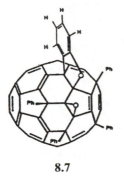

8.7

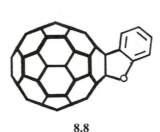

8.8

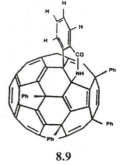

8.9

It is the presence of a 6:5-double bond in $C_{60}Ph_4O_2$ which permits elimination to give $C_{58}Ph_4$, the conjectured structure of which is **8.6**. Similar argument account for the formation of $C_{68}Ph_4$ described below (Sec. 8.6.2).

The major component produced by oxidation of $C_{60}Ph_5H$ is a benzo[*b*]furan derivative **8.7**, which is obtained also by similar treatment of the $C_{60}Ph_5Cl$ precursor. The mechanism of formation is not known, but the hydride precursor may form a hydroperoxide which then eliminates water.[31]

The *parent* compound of **8.7**, *viz.* benzo[*b*]furano[2,3:1,2][60]fullerene **8.8**, can be made simply by reaction of $C_{60}Cl_6$ with phenol in the presence of aq. KOH.[32] The mechanism of this reaction is not yet known, but must involve nucleophilic substitution at some stage. It is one of many novel reactions that are being uncovered in fullerene chemistry.

Another surprising reaction is that between $C_{60}Ph_5Cl$ and $BrCN/FeCl_3$ which results in formation of the phenylated isoquinolono[3,4:1,2][60]fullerene **8.9**, arising from nucleophilic substitution of Cl by BrCN, followed by electrophilic substitution of the cyano carbon into the *ortho* position of the adjacent phenyl ring, and hydrolysis of the C-Br bond.[33]

8.6.2 *Substitutions in* $C_{70}Cl_{10}$

Reaction of $C_{70}Cl_{10}$ with benzene and ferric chloride produces two main products, $C_{70}Ph_{10}$, **8.10** and $C_{70}Ph_8$, **8.11**. An interesting mechanistic feature is that **8.10** (7,8,19,26,33,37,45,49,53,63-decaphenyl-7,8,19, 26,33,37,45,49,53,63-decahydro[60]fullerene seems not to be produced directly from the chloro precursor, but rather by further phenylation of **8.11** (7,19,23,27,33,37,44,53-octaphenyl-7,19,23,27,33,37,44, 53-octahydro[70]fullerene.[34] This is indicated both by formation of the octaphenyl compound being the more rapid, and the fact that two tolyl groups can be added to the octaphenyl compound by reaction with toluene/ferric chloride. Formation of the decaphenyl compound must therefore involve Friedel-Crafts alkenylation in the last steps. Steric hindrance may contribute to the failure to get direct replacement of the last two chlorines; rotation of the adjacent phenyl groups in **8.10** is sterically restricted. A driving force for the alkenylation step is that it takes place across the only double bond in $C_{70}Ph_8$ that is in a pentagonal ring. This introduces strain which is partly relieved on addition across the bond. For the same reason, a [4 + 2] cycloaddition with anthracene occurs readily across this bond.[35]

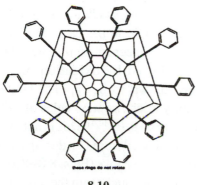

8.10

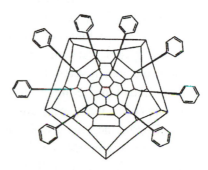

8.11

| 8.12 | 8.13 | 8.14 |

A byproduct of the phenylation reaction is $C_{70}Ph_9OH$, **8.12**, the first monohydroxyfullerene to be prepared, and here the adjacency OH and phenyl groups results in restricted rotation of the latter.[36] The mechanism of formation of this compound is obtuse, given that $C_{70}Ph_{10}$ appears not to be obtained directly from the decachloro precursor. Two isomeric and symmetrical diols, $C_{70}Ph_8(OH)_2$ are also obtained from reacting a dichloromethane solution of $C_{70}Ph_8$ with 18-crown-6 and $KMnO_4$. These are 19,26,33,37,45,49,53, 63-octaphenyl-7,8,19,26,33,37,45,49,53,63-decahydro[70]fulleren-7,8-diol, **8.13** and 19,26,33,37,45,49,53,63-octaphenyl-19,22,23,26,33,37,45,49,53, 63-decahydro[70]fulleren-22,23-diol, **8.14**. The former is also obtained by reacting a CCl_4 solution of $C_{70}Ph_8$ in air in the presence of [70]fullerene (presumed to act as singlet oxygen producer).[37] These are the first 1,2-diols derived from [70]fullerene, and like the fullerenols described above are very insoluble in many organic solvents. A 1,2-diol derived from a cyclopropanated [60]fullerene has also been made by the permanganate oxidation, but differs from the above in being amenable to oxidation to a dioxetane.[38]

On standing in air, $C_{70}Ph_8$ undergoes a ring-opening oxidation to give the *bis* lactone **8.15** which has an eleven-membered hole in the cage, the least restricted opening yet achieved in a fullerene.[39] The mechanism is believed to involve oxygen insertion into 6:5-bonds followed by oxidation of the intermediate vinyl ether, a general process known to be rapid.[40] Under conditions of EI mass spectrometry, **8.15** readily loses two molecules of CO_2 to give $C_{68}Ph_8$, the structure of which is believed to be **8.16**.[30,41] This is notable, because only minor traces of C_{68} are normally produced from C_{70} during mass spectrometry. The reason follows the explanation given above (Figs. 8.8–8.10). $C_{70}Ph_8$ has the required 6:5-double bond, removal of which from the tetroxide

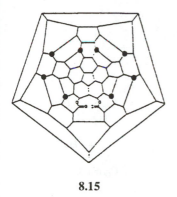

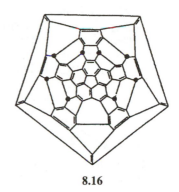

8.15	8.16

(in this case as $2 \times CO_2$) gives $C_{68}Ph_8$. Moreover, although there are two *adjacent* pentagons in the structure, two are next to heptagonal ring which removes the strain, and this is diminished further by the sp^3-hybridised carbons common to each pair. The unique structure of $C_{70}Ph_8$ therefore facilitates ring contraction. In general the presence of addends on the cages is likely to improve the prospects of isolating irregular fullerenes; the possibility of fullerenes having 7-membered rings was predicted somewhat earlier,[42] and has been supported recently by calculations.[43]

8.6.3 Substitution in $C_{60}F_{18}$

Of the halogenofullerenes, fluorofullerenes are the most reactive towards nucleophilic substitution, and they are also much the most soluble. They react readily with water to replace F by OH in a process which probably involves addition-elimination. Scheme 8.1 shows the species (including fragmentation ions) that have been isolated and characterised by mass spectrometry, arising from the reaction of $C_{60}F_{36}$ with water.[44] Similar species have been isolated from reaction of $C_{70}F_{34/36/38/40}$ with water, and in some cases characterised by ^{19}F NMR.[45]

    $C_{60}F_{18}$ will react with benzene and ferric chloride merely on standing for a few days at room temperature., but compounds with a higher fluorination level are more resistant, due presumably to more restricted access to the reaction sites. The product of this reaction is the remarkable *triumphene* ($C_{60}F_{15}Ph_3$), which has three phenyl groups in a trefoil-shaped arrangement **8.17**. $C_{60}Ph_{18}$

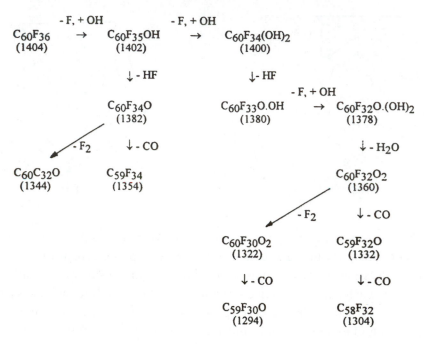

$$
\begin{array}{ccccc}
\overset{\text{- F, + OH}}{C_{60}F_{36}} & \rightarrow & \overset{\text{- F, + OH}}{C_{60}F_{35}OH} & \rightarrow & C_{60}F_{34}(OH)_2 \\
(1404) & & (1402) & & (1400)
\end{array}
$$

$\downarrow$ - HF $\downarrow$ - HF

$C_{60}F_{34}O$ $C_{60}F_{33}O.OH$ $\xrightarrow{\text{- F, + OH}}$ $C_{60}F_{32}O.(OH)_2$

(1382)   (1380)   (1378)

- F$_2$ $\downarrow$ - CO $\downarrow$ - H$_2$O

$C_{60}C_{32}O$ $C_{59}F_{34}$ $C_{60}F_{32}O_2$

(1344)   (1354)   (1360)

- F$_2$ $\downarrow$ - CO

$C_{60}F_{30}O_2$ $C_{59}F_{32}O$

(1322)   (1332)

$\downarrow$ - CO $\downarrow$ - CO

$C_{59}F_{30}O$ $C_{58}F_{32}$

(1294)   (1304)

Scheme 8.1 Products from nucleophilic substitution by water, of F in $C_{60}F_{36}$.

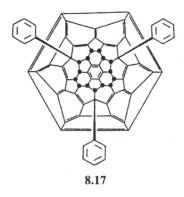

8.17

has also been isolated from this reaction, and this is not only the most phenylated fullerene (or indeed any other compound) yet obtained, but most notably, the phenyl group occupy the same sites as the fluorines in the precursor, so that each phenyl group is adjacent to at least one other.[46]

8.6.4 Substitutions in Bromo[60]fullerenes

The first report of this reaction was by Cooks and coworkers who found that whereas up to six aryl rings became attached to [60]fullerene if it was heated with ferric chloride and aromatics such as toluene, *p*-xylene, or anisole, the reaction would only occur with benzene if bromine was present.[47] This can now be seen to be due to the lower electrophilic reactivity of benzene which will only react with the more reactive halogenated fullerene electrophile. The reaction is less controlled than those above using prepurified halogenofullerenes, but with the advent of improved HPLC methods of separating fullerene derivatives, there lies the prospect of being able to separate and characterise many more compounds by this route.

Compounds which have been identified (and in some case separated by HPLC) from the reaction of [60]fullerene with bromine, benzene and ferric chloride are as follows:[30,48] $C_{60}Ph_n$ ($n = 4$, 6, 8, 10, 12), $C_{60}Ph_nO_2$ ($n = 4,6,8,10,12$), $C_{60}Ph_8O_4$, $C_{60}Ph_nOH$ ($n = 7$, 9, 11), $C_{60}Ph_nH_2$ ($n = 4$, 10), $C_{60}Ph_4H_4$, $C_{60}Ph_5H_3$, $C_{60}Ph_nO_2H$ ($n = 5, 9$), $C_{60}Ph_4C_6H_4O_2$ (**8.7**), $C_{60}Ph_9OH_3$, and $C_{60}Ph_{11}O_3H_2$. In the corresponding reaction with toluene and chlorobenzene, $C_{60}(MeC_6H_4)_4$ and $C_{60}(ClC_6H_4)_5H$ are the main products, respectively.

Just as $C_{60}Ph_4O_2$ loses two molecules of CO during mass spectrometry [to give $C_{58}Ph_n$ derivatives (Sec. 8.6.1)], so too does $C_{60}Ph_8O_2$. By contrast $C_{60}Ph_8O_4$ does *not* degrade in the same way, because this would produce a C_{56} derivative which is presumably too unstable.

8.7 Fullerenylation

This reaction consists of the electrophilic substitution of the fullerene into aromatics in the presence of Lewes acid catalysts. Aromatics PhR, where R = H, Me, 1,3-Me$_2$, F, Cl, OMe and NMe$_2$, have been studied in the reaction with [60]fullerene,[47] and the effectiveness of the catalysts is: AlBr$_3$ > AlCl$_3$ > FeCl$_3$ > FeBr$_3$ > GaCl$_3$ > SbCl$_5$,[45,47] whilst TiCl$_4$, SnCl$_4$, GeCl$_4$, BF$_3$ and BCl$_3$ are ineffective.[49] The reaction is faster with the more electron-rich aromatics, showing it to be an electrophilic aromatic substitution. The proposed mechanism, *viz.* protonation of the fullerene by residual protons in the catalyst to give $C_{60}H^+$ as an electrophile,[47] is analogous to that considered

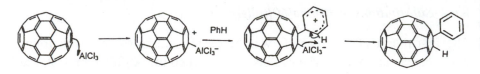

Fig. 8.11 Conjectured mechanism for Friedel-Crafts fullerenylation of aromatics.

to apply to electrophilic aromatic alkylation by alkenes in the presence of FC catalysts.[50] However, given that it is not possible to protonate fullerenes with superacids, though radical cations have been observed,[51] an alternative mechanism may apply. This involves polarisation of a fullerene π-bond by the catalyst, thereby creating an electrophilic centre which then attacks the aromatic (Fig. 8.11).

This FC-catalysed alkylation of aromatics is faster with [60]fullerene than [70]fullerene.[3]

References

1. G. A. Olah, *Friedel Crafts and Related Reactions*, Interscience, 1963, vol. 1, p. 40.
2. R. Taylor *et al.*, *J. Chem. Soc., Chem. Commun.*, (1994) 275.
3. A. G. Avent, P. R. Birkett and R. Taylor, unpublished work.
4. G. A. Olah *et al.*, *J. Am. Chem. Soc.*, **113** (1991) 9385.
5. R. Taylor *et al.*, *J. Chem. Soc., Chem. Commun.*, (1992) 665.
6. P. R. Birkett, A. D. Darwish, H. W. Kroto, R. Taylor, D. R. M. Walton and O. B. Woodhouse, unpublished work.
7. R. Taylor, *Electrophilic Aromatic Substitution*, Wiley, Chichester, 1989.
8. L. Y. Chiang, J. W. Swirczewski, C. S. Hsu, S. K. Chowdhury, S. Cameron and K. Creegan, *J. Chem. Soc., Chem. Commun.*, (1992) 1791; L. Y. Chiang, R. B. Upasani and J. Swirczewski, *J. Am. Chem. Soc.*, **114** (1992) 10154.
9. L. Y. Chiang, L. Y. Wang, J. W. Swirczewski, G. Miller, S. Soled and S. Cameron, *J. Org. Chem.*, **59** (1994) 3960.
10. J. H. Gorvin, *J. Chem. Soc.*, (1959) 678; (1963) 3980.
11. A. Hamwi and V. Marchand, *Fullerene Sci. & Technol.*, **4** (1996) 835.

12. L. Y. Chiang, R. B. Upasani, J. W. Swirczewski and S. Sold, *J. Am. Chem. Soc.*, **115** (1993) 5453.
13. Ref. 7, pp. 270–272.
14. T. Pradeep and R. G. Cooks, *Int. J. Mass Spectrom., Ion Procs.*, **135** (1994) 243; P. R. Birkett *et al.*, *J. Chem. Soc., Perkin Trans. 2*, (1997) 1121.
15. G. P. Miller, M. A. Buretea, M. M. Bernardo, C. S. Hsu and H. L. Fang, *J. Chem. Soc., Chem. Commun.*, (1994) 1549.
16. S. Roy and S. Sakar, *J. Chem. Soc., Chem. Commun.*, (1994) 275.
17. L. Y. Chiang, J. B. Bhonsie, L. Wang, S. F. Shu, T. M. Chang and J. R. Hwu, *Tetrahedron*, **52** (1996) 4963.
18. N. S. Schneider, A. D. Darwish, H. W. Kroto, R. Taylor and D. R. M. Walton, *J. Chem. Soc., Chem. Commun.*, (1994) 463.
19. J. Li, A. Takeuchi, M. Ozawa, X. Li, K. Saigo and K. Kitazawa, *J. Chem. Soc., Chem. Commun.*, (1993) 1784.
20. L. Y. Chiang, F. Lu and J. Lin, *J. Chem. Soc., Chem. Commun.*, (1995) 1283.
21. C. Yu, J. B. Bhonslee, L. Y. Wang, J. G. Liu, B. Chen and L. Y. Chiang, *Fullerene Sci. & Technol.*, **5** (1997) 1407.
22. L. Y. Chiang, F. Lu and J. Lin, *Proc. Electrochem. Soc.*, **95–10** (1995) 699.
23. H. Huang, L. Lu and L. Y. Chiang, *Proc. Electrochem. Soc.*, **95–10** (1996) 403.
24. E. Nakamura, H. Tokuyama, S. Yamago, T. Shiraka and Y. Sugiura, *Bull. Chem. Soc. Jpn.*, **69** (1996) 2143.
25. A. K. Abdul-Sada, A. G. Avent, P. R. Birkett, H. W. Kroto, R. Taylor and D. R. M. Walton, *J. Chem. Soc., Perkin Trans. 2*, (1998) 393.
26. A. G. Avent *et al.*, *J. Chem. Soc., Chem. Commun.*, (1994) 1463.
27. P. R. Birkett *et al.*, *J. Chem. Soc., Perkin Trans. 2*, (1997) 1121.
28. R. Taylor, *Electrophilic Aromatic Substitution*, Wiley, Chichester, 1989, Chap. 11.
29. P. R. Birkett, A. G. Avent, A. D. Darwish, H. W. Kroto, R. Taylor and D. R. M. Walton, *J. Chem. Soc., Perkin Trans. 2*, (1997) 457.
30. A. D. Darwish, P. R. Birkett, G. J. Langley, H. W. Kroto, R. Taylor and D. R. M. Walton, *Fullerene Sci. & Technol.*, **5** (1997) 705.
31. A. G. Avent, P. R. Birkett, A. D. Darwish, H. W. Kroto, R. Taylor and D. R. M. Walton, *Chem. Commun.*, (1997) 1579.

32. A. D. Darwish and R. Taylor, unpublished work.
33. A. K. Abdul-Sada *et al.*, *Chem. Commun.*, (1998) 307. .
34. A. G. Avent, P. R. Birkett, A. D. Darwish, H. W. Kroto, R. Taylor and D. R. M. Walton, *Tetrahedron*, **52** (1996) 5235.
35. A. G. Avent, P. R. Birkett, A. D. Darwish, H. W. Kroto, R. Taylor and D. R. M. Walton, *Fullerene Sci. & Technol.*, **5** (1997) 643.
36. P. R. Birkett, A. G. Avent, A. D. Darwish, H. W. Kroto, R. Taylor and D. R. M. Walton, *J. Chem. Soc.*, *Chem. Commun.*, (1996) 1239.
37. P. R. Birkett, A. G. Avent, A. D. Darwish, H. W. Kroto, R. Taylor and D. R. M. Walton, *J. Chem. Soc.*, *Perkin Trans. 2*, unpublished work.
38. I. Lamparth, A. Herzog and A. Hirsch, *Tetrahedron*, **52** (1996) 5065.
39. P. R. Birkett, A. G. Avent, A. D. Darwish, H. W. Kroto, R. Taylor and D. R. M. Walton, *J. Chem. Soc.*, *Chem. Commun.*, (1995) 1869.
40. R. Taylor, *J. Chem. Res. (S)*, (1987) 178.
41. P. R. Birkett, A. D. Darwish, A. G. Avent, H. W. Kroto, R. Taylor and D. R. M. Walton, *The Chemical Physics of Fullerenes 10 (and 5) Years Later* (ed. W. Andreoni), NATO ASI Series E: Applied Sciences, **316** (1996) 199.
42. R. Taylor, *Interdisciplinary Sci. Rev.*, **17** (1992) 171.
43. A. Ayuela, P. W. Fowler, D. Mitchell, R. Schmidt, G. Seifert and F. Zerbetto, *J. Phys. Chem.*, **100** (1996) 15634.
44. R. Taylor *et al.*, *J. Chem. Soc.*, *Perkin Trans. 2*, in press.
45. O. V. Boltalina, J. M. Street and R. Taylor, unpublished work.
46. O. V. Boltalina, J. M. Street and R. Taylor, *Chem. Commun.*, in press.
47. S. H. Hoke, J. Molstad, G. L. Payne, B. Kahr, D. Ben-Amotz and R. G. Cooks, *Rapid Comm. Mass Spectr.*, **5** (1991) 472.
48. R. Taylor *et al.*, *J. Chem. Soc.*, *Chem. Commun.*, (1992) 667.
49. G. A. Olah, *et al.*, *J. Am. Chem. Soc.*, **113** (1991) 9387; G. A. Olah, I. Bucsi, D. S. Ha, R. Anisfeld, C. S. Lee and G. K. S. Prakash, *Fullerene Sci. & Technol.*, **5** (1997) 389.
50. Ref. 28, p. 202.
51. G. A. Olah, I. Bucsi, R. Anisfeld and G. K. S. Prakash, *Carbon*, **30** (1992) 1203.

9

Cycloadditions

Of all fullerene reactions, cycloadditions have received by far the most study. This popularity stems from the ability to control the reaction so that only one addend becomes attached to the cage, making analysis of the produces relatively easy. The object of many of these syntheses is to produce intermediates for further reaction, though some of the functional groups do not undergo their normal reactions either readily or at all, when attached to fullerenes. This is due to both the strong electron withdrawal by the cage, and steric constraints.

Six types of reaction are known: [1 + 2], [2 + 2], [3 + 2], [4 + 2], [6 + 2] and [8 + 2] cycloadditions. The [4 + 2] group (Diels-Alder reactions) have been the subject of most attention. The number of compounds that have been made by cycloaddition is already too vast to be described in detail here, consequently the reactions shown are selected either to be representative of general features, or have some aspect of special interest. There have been recent reviews on some of these reactions.[1-3]

9.1 [1 + 2] Cycloadditions: Reactions That Produce Methano- and Homofullerenes and Their Heteroanalogues

These additions involve either carbon, oxygen, nitrogen and silicon, numerous derivatives based upon these additions being now known. Addition of carbon has been the most studied to date, but addition of nitrogen (which leads to azafullerenes described in Chap. 13) and of oxygen (which can occur spontaneously) are rapidly gaining in importance. Two possible products are obtainable, arising from insertion into a 6,5 σ-bond giving **9.1**, or addition to a

9.1   9.2

6,6 π-bond giving **9.2**, the reason for these preferences having been given in Sec. 3.1. For a single atom like oxygen, only one product is obtainable in each case, but for the addition of either > NR or > CRR′ then two products are possible from **9.1**, depending upon the conformation of the addend with respect to the 5- and 6-membered rings.

9.1.1 *Addition of Carbon*

9.1.1.1 *Additions Involving Carbenes and Azo Precursors*

Precursors that have been used include diazo compounds (including diazirines, oxadiazoles and lithium salts of tosylhydrazones), α,α,α trihalocarboxylates and cyclopropenone acetals. The simplest addition is that of methylene which can give either 1(6)a-homo[60]fullerene (**9.1**, $X = CH_2$) or 1,2-methano[60]fullerene (**9.2**, $X = CH_2$).

Methylenation evidently occurs readily, since species of $M + 14$, $M + 28$, $M + 42$ etc. amu often appear in the mass spectra of fullerenes and derivatives. This was first observed in soot prepared by the arc-discharge method, giving derivatives of both [60]- and [70]fullerenes.[4] The products of reaction of bromofullerenes with THF and tetrahydropyran contained both methylene and polymethylene derivatives of [60]-, [70]-, [78]- and [84]fullerenes; the polymethylene compounds appear to arise from addition of $(CH_2)_n$, $n = 3,4$, moieties, rather than multiple addition of methylene groups.[5] Methylene derivatives also accompany fullerene hydrogenation,[6] due most probably to fragmentation of some of the cages giving the carbene: CH_2 which then adds to other cages. Likewise, addition of CF_2 accompanies fluorination of both

[60]- and [70]fullerenes, again attributable to cage degradation producing the corresponding difluorocarbene.[7]

Controlled addition of methylene and substituted methylene is achieved very effectively through the use of diazo compounds. These may either eliminate nitrogen to give a carbene (which then participates in [1 + 2] cycloaddition to the cage), or they can carry out [3 + 2] cycloaddition to give an intermediate pyrazoline. On heating this loses nitrogen to give almost exclusively the homo[60]fullerene 6,5-insertion product (also known as a fulleroid), whilst on irradiation it gives both this and the methano[60]fullerene 6,6-addition product, Fig. 9.1.[8-11] The former has C_s symmetry, the ^{13}C NMR spectrum showing 32 sp^2 signals for the cage $[(28 \times 2 \text{ C}) + (4 \times 1 \text{ C})]$, and two doublets in the ^{1}H NMR (δ 2.87 and 6.35, the latter downfield resonance being due to the hydrogen lying over the π-deficient pentagon).The methano[60]fullerene has C_{2v} symmetry, the ^{13}C NMR showing 16 sp^2 signals $[(13 \times 4 \text{ C}) + (3 \times 2 \text{ C})]$ together with one sp^3 signal at δ 71.0 (2 C) for the cage carbons, and a ^{1}H NMR singlet at δ 3.93. Similar analyses distinguish the two types of compounds in various derivatives, plus the fact that the methano[60]fullerenes generally show a peak in the UV-VIS at *ca.* 434 nm. X-ray studies on compounds with both CAr$_2$ and C(C $\equiv$ C.C $\equiv$ CSiMe$_3$)$_2$ bridges, confirm the presence of the cyclopropane ring in the methanofullerenes.[12]

Whilst the parent compounds shown in Fig. 9.1 do not rearrange to each other,[11] the 6,5-(open cage) compounds having substituted methylene bridges rearrange either thermally,[13] photochemically,[14] electrochemically[15] or under acid catalysis,[16] to the 6,6-methanofullerenes, which are thermodynamically more stable[17,18] (though for a recent exception, see below). This process

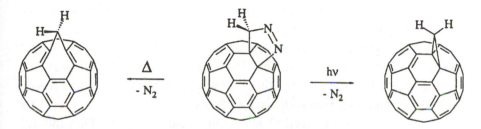

Fig. 9.1 Thermal and photochemical decomposition of a pyrazoline[60]fullerene to give a 1(6)a-homo[60]fullerene or a 1,2-methano[60]fullerene, respectively.

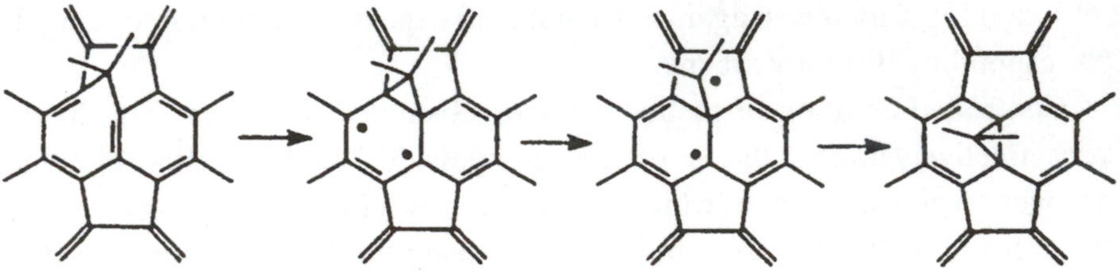

Fig. 9.2 Postulated mechanism for the 6,5-homo- to 6,6-methano-fullerene rearrangement.

probably involves an electrocyclisation to give the 6,5-(closed cage) compound followed by a [1,5] shift (Fig. 9.2).[9] However, the thermal process is zeroth order, being catalysed by light which promotes formation of a triplet bidiradical intermediate in the electrocyclisation step. This also explains the strong rearrangement-inhibiting effect of oxygen.[19]

The absence of any 6,6-open-cage compound is due to the strong destabilising effect that would arise from the necessary placement of three double bonds in pentagons (Sec. 3.1). This contrasts with the closely related 1,6-methano[10]annulenes, for which the open structure (no bonding between the 1- and 6-positions) is favoured, except when strongly electron-withdrawing substituents are on the methylene bridge.[20] Mutual repulsion by these increases the bond angle between them and reduces it between the other two bonds, which brings the 1- and 6-positions closer together, thus facilitating σ-bonding. For the fullerenes, however, the corresponding bonding occurs regardless of substituents on the methylene group.

Numerous 1,2-methano[60]fullerenes have been made with substituents (especially aryl groups) on the methylene group CRR'. Combinations of RR' include H,Ar and Me,Ar and in the latter case both homo[60]fullerene isomers are also obtained,[15,17] as well as H,CO_2R" (from diazoacetate precursors)[21] and H,CONR"$_2$ (from diazoamide precursors),[22] and HC $\equiv$ C,C $\equiv$ CSi(i-Pr)$_3$.[23] Whereas carboxyalkyl groups attached to the cage are difficult to hydrolyse to the free acid, this is easier when these groups are substituted in the bridging methylene group. Reaction with NaH achieves the required conversion,[24] and the acids can then be converted into amides etc.[24,25] 61,61-Dicyano-1,2-methano[60]fullerene [$C_{60}C(CN)_2$] is notable for the exceptional electron withdrawal by the addend.[26] Thus the first reduction potential occurs at a value

156 mV more positive than for [60]fullerene, whereas in methano[60]fullerenes the first three reduction waves are usually each 100–150 mV more negative. This suggests that there is through-space (orbital interaction) electron withdrawal (*periconjugation* — see below) between the methano substituents and the cage.

A range of methano[60]fullerenes with CAr_2 bridges are known in which the *p*-substituent in the aryl rings are H, Me, OMe, Br, NMe_2, CO_2Ph, OH and $(CH_2)_2NHCOMe$,[10] as are derivatives in which the aryl group in the methylene group CPhAr is either a [15]crown-5-,[27] or [18]crown-6-substituted phenyl.[28] Immersion of a gold surface modified by a thiol-terminated ammonium salt into a dichloromethane solution of the latter, results in the (reversible) attachment of a monolayer of the [60]fullerene derivative **9.3**. A methano[60]fullerene cryptate, prepared by reaction of an amine precursor with 1,2-methano[60]fullerene-61-carboxylic acid, involves binding of a sodium ion to the benzo[2,2,2]cryptand moiety **9.4**.[29]

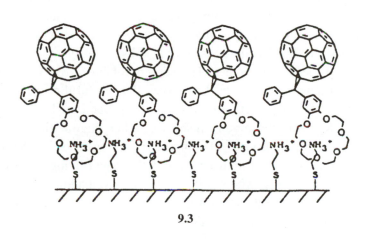

9.3

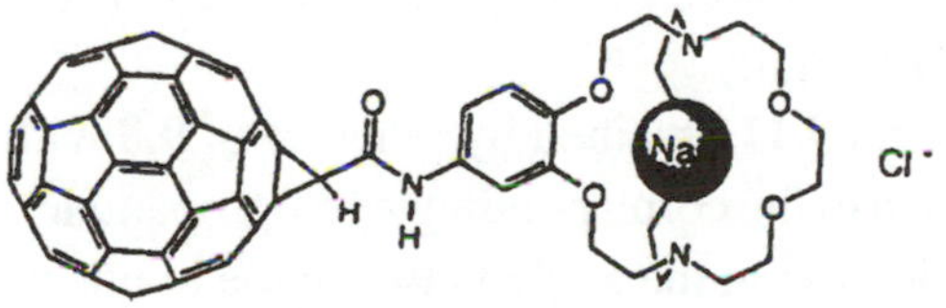

9.4

A diarylmethano[60]fullerenes of particular interest is the *bis* acid **9.5**. Modelling studies indicated that the [60]fullerene cage would fit into the most hydrophobic active site of the enzyme HIV-1 protease,[30] so water-soluble **9.5** was synthesised for clinical tests, and produced a promising inhibiting effect on chronically HIV-1 infected human peripheral blood mononuclear cells.[31] This result has stimulated the synthesis of methano[60]fullerenes with biochemical potential, examples having amino acid groups present on the addend, e.g. **9.6**.[32]

An amphiphilic derivative with R = H, R′ = $(CH_2)_9CO_2Et$ is unusual in being a homofullerene which, moreover, does not rearrange to the methanofullerene on heating, currently a unique situation.[33]

The methano[60]fullerenes have also been used to investigate model structures for fullerene-containing polymers (see Sec. 12.2, Figs. 12.8 and 12.9).

Other examples of precursors are **9.7**, **9.8**[25,34] and **9.9**[35], each of which gives the corresponding methano[60]fullerene derivative, the former producing a fullerene sugar; the carbene is formed from **9.8** by loss of both nitrogen and acetone, and from **9.9** by loss of nitrogen and the lithium *p*-toluene sulphonate. In the formation of spiranes using the latter reagent type, short reaction times yield mainly the homofullerenes which subsequently rearrange to the corresponding methanofullerenes. The carbene obtained by N_2 and sodium p-toluenesulphonate loss from **9.10** gives the homofullerene,[36] and the carbenes: CR_2 [$R_2 = Me_2$, Et_2, Pr_2, Bu_2, $(CH_2)_5$], obtained by nitrogen loss from $N_2 = CR_2$ also give mainly insertion rather than addition (the cage acquiring up to two addends).[37] Timing of the nitrogen loss is the deciding factor, for if this occurs before reaction with the cage (giving a free carbene) then 6,6-addition results, whereas if 1,3-dipolar cycloaddition of the diazo intermediate occurs and is then followed by nitrogen loss, 6,5-insertion results.[37] Similar insertion occurs if phenylchlorodiazirine is used as precursor,[38] and a fascinating aspect here is that up to seven CArCl groups add to the cage (nine detected in a mass spectrometry study)[39] the structure of which must therefore be altered very substantially.

The dimethyl ketal **9.11** obtained by reaction of **9.8** with [60]fullerene[25,40] cannot be hydrolysed to the corresponding cyclopropanone, which might seem to arise from the increase in the angle between the bonds to the fullerene, and hence strain. Instead **9.11** produces 1-methylcarboxy-1,2-dihydro[60]fullerene which resists saponification by sulphuric acid, boron trifluoride and

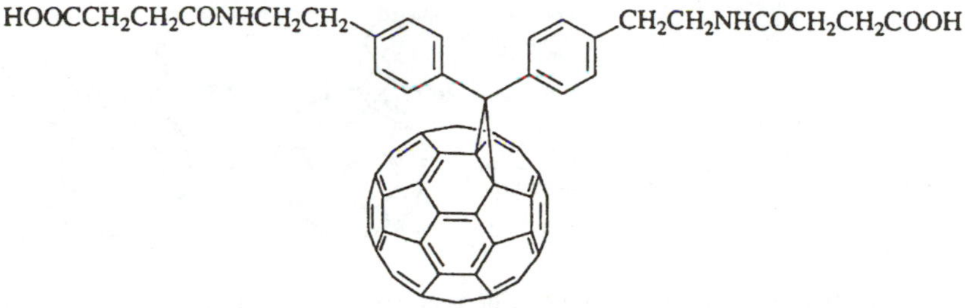
HOOCCH2CH2CONHCH2CH2
CH2CH2NHCOCH2CH2COOH

9.5

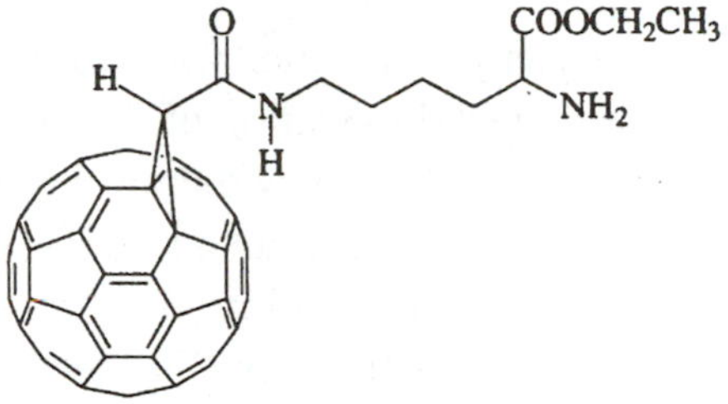
O
COOCH2CH3
H
N
H
NH2

9.6

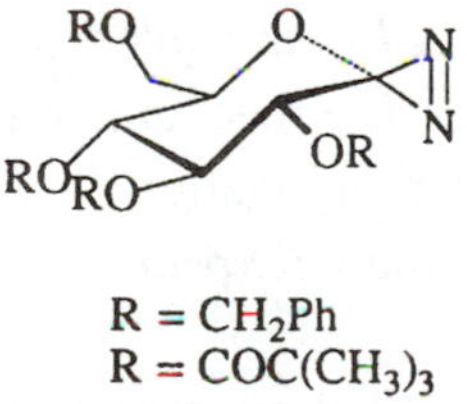
RO
O
N
N
OR
RO
RO
R = CH2Ph
R = COC(CH3)3

9.7

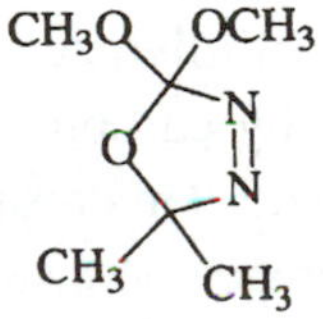
CH3O
OCH3
N
N
O
CH3
CH3

9.8

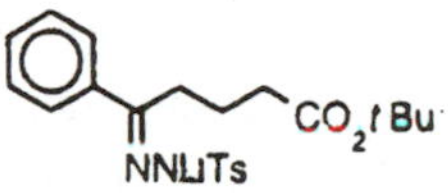
CO2tBu
NNLiTs

9.9

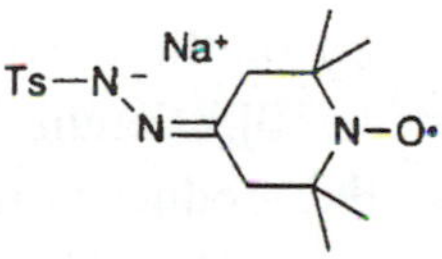
Na+
Ts—N-
N=
N–O•

9.10

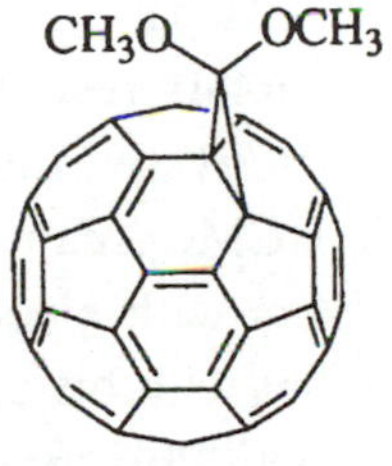
CH3O
OCH3

9.11

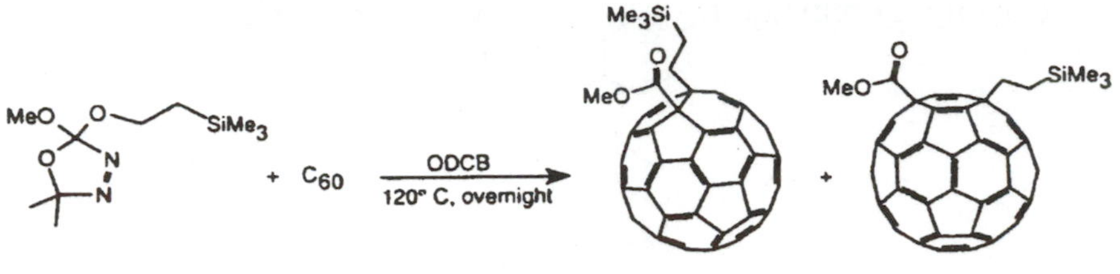

Fig. 9.3 Formation of dihydro[60]fullerene derivatives by reaction of [60]fullerene with a 1,3,4-oxadiazoline.

trimethylsilyl iodide. Phase-catalysed reaction with HBr produces [60]fullerene only, almost certainly due to decarboxylation of the intermediate carboxylic acid (such processes are aided by electron withdrawal so would be very facile here) and subsequent degradation of the resultant $C_{60}H_2$ (known to be unstable — see Chap. 4). (Formation of the cyclopropanone ring by hydrolysis of 1′,1′-dichloro-1,2-methano[60]fullerene has been achieved,[41] so an explanation other than that based upon strain appears to be necessary.) Attempted addition of a alkoxy carbene to [60]fullerene through thermolysis of a 2,2-dialkoxy-Δ^3-1,3,4-oxadiazoline precursor produces none of the expected product, but instead a mixture of 1,2- and 1,4-(methoxycarbonyl)-[(trimethylsilyl)ethyl]dihydro[60]fullerenes (Fig. 9.3);[42] the mechanism may involve rearrangement (driven by carbon-silicon hyperconjugation[43]) of an initially-formed methanofullerene. This result typifies the unexpected that is encountered frequently in fullerene chemistry.

The 1′,1′-dichloro-1,2-methano[60]fullerene described above was formed by pyrolysis of sodium chloroacetate (which gives: CCl_2) in the presence of the fullerene.[41] It can also been made by generating the carbene from chloroform in the presence of a base.[38] Elimination of Hhal by base from the corresponding haloform precursors results in the addition of CBr_2, CBrCl, CI_2 (the only iodofullerene derivative) and CHCN.[44] The product of the latter addition, and also $C_{61}HCMe_3$, can be produced by reaction of the corresponding dihaloprecursors with electrochemically generated [60]fullerene dianion.[45] $C_{61}Br_2$ (together with bis and tris adducts) is also the product of reaction of [60]fullerene with $PhHgBr_3$, and is a precursor for formation of C_{121} and C_{122} (see Sec. 12.6).[46]

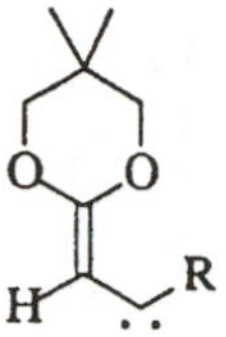

9.12

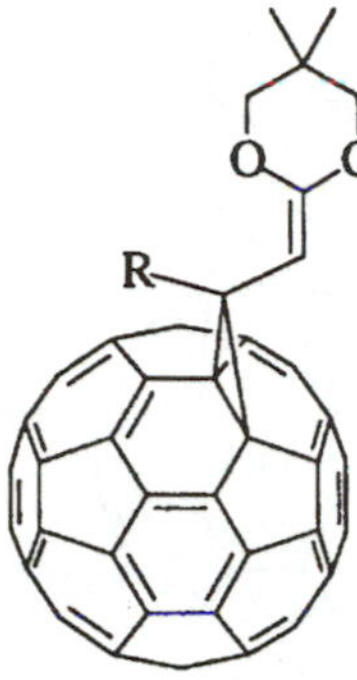

9.13

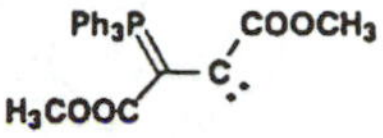

9.14

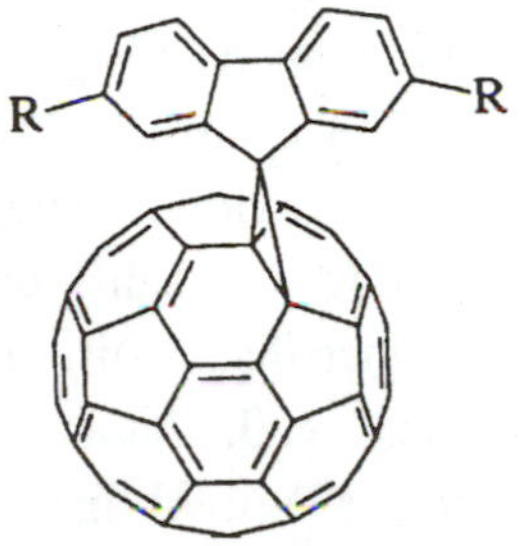

9.15 [R = H; N(CH$_2$Ph)$_2$; NO$_2$]

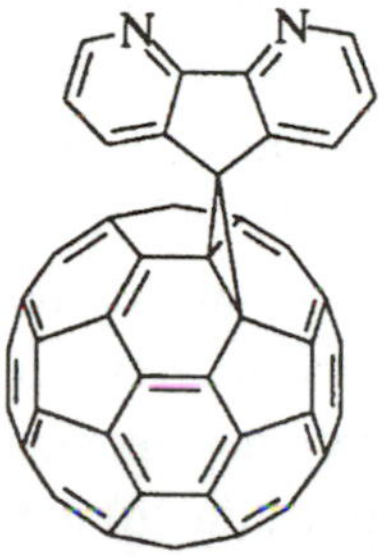

9.16

Methano[60]fullerenes can be obtained by reaction of the fullerene with a vinylcarbene (e.g. **9.12** from thermolysis of cyclopropene acetals) giving **9.13** (which is readily hydrolysable to ester);[18] this reaction is accompanied by [3 + 2] addition products, the proportion of the latter increasing with increasing reaction temperature. They are also obtained from the reaction of dimethyl acetylenedicarboxylate (butyn-1,4-dioate) in the presence of triphenylphosphine, which produces the intermediate carbene-ylid (**9.14**).[47]

The electronic properties of [60]fullerene are not significantly altered in either methano- or homofullerenes, and the effect of substituents in aryl groups attached to the methylene group are trivial as might be expected since there is no possibility of conjugation between the aryl groups and the cage. This is not the case however for the spiranes **9.15** and **9.16**, for here the aryl groups are necessarily orthogonal to the cage and even point into the cage surface.

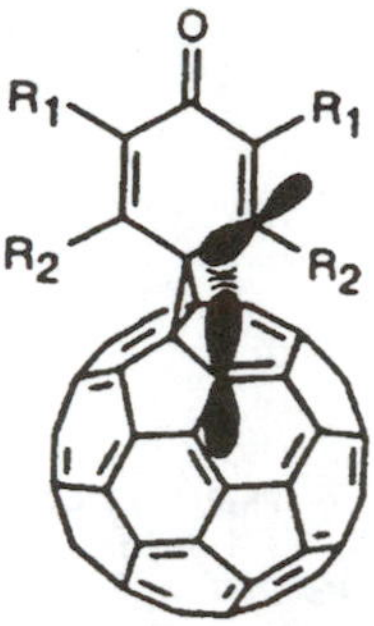

9.17    **9.18** 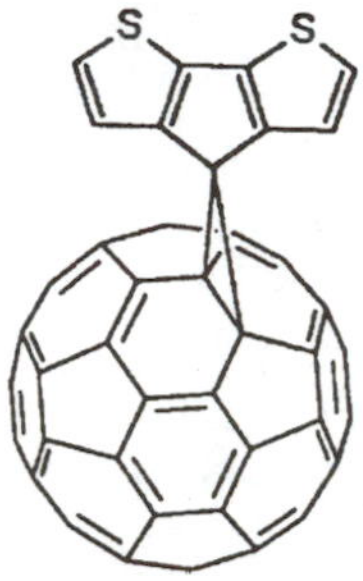   **9.19**

Electronic interactions therefore occur (named 'periconjugation') and as might be expected, electron-donating substituents [such as $N(CH_2Ph)_2$] cause a shift in the reduction potentials (for addition of up to five electrons, measured by cyclic voltammetry) to more negative values i.e. the molecule becomes harder to reduce, and vice versa. Thus the 4,4'-diazafluoren-9-yl-1,2-methano[60]fullerene **9.16** is actually easier to reduce than [60]fullerene itself.[48]

Likewise quinone-type methano[60]fullerenes e.g. **9.17** are easier to reduce than [60]fullerene, due to periconjugation (see **9.17**) involving the p-orbitals of the cyclohexadienone ring and those of the cage adjacent to the cyclopropane bridge. However, when bulky substituents R_2 are present in **9.17**, and for the anthrone derivative **9.18**, the electron withdrawal properties are reduced due to steric inhibition of the periconjugation.[49] Cyclic voltammetry of a spirane with [60]fullerene linked to a cyclopentadithiophene group **9.19** shows that the two moieties retain their individual electrochemical properties.[50] Here the greater distance between the relevant p-orbitals, probably reduces the opportunity for periconjugation.

Multiple addition levels all the way up to the hexa-adduct (which must have an octahedral addition pattern) are found for the addition of CPh_2 (from excess of the diazo precursor) to [60]fullerene, but the products are as yet uncharacterised;[8] the bimolecular rate coefficient for the monoaddition is 4.0×10^8 M^{-1} s^{-1}.[51] More extensive studies of polyaddition are described in Sec. 9.1.1.5.

All of the above spiranes are methanofullerenes, but one homofullerene spirane is produced by reaction of [60]fullerene with a mixture of

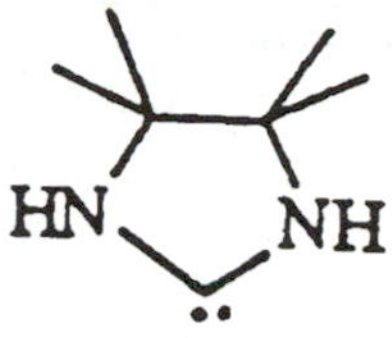

9.20

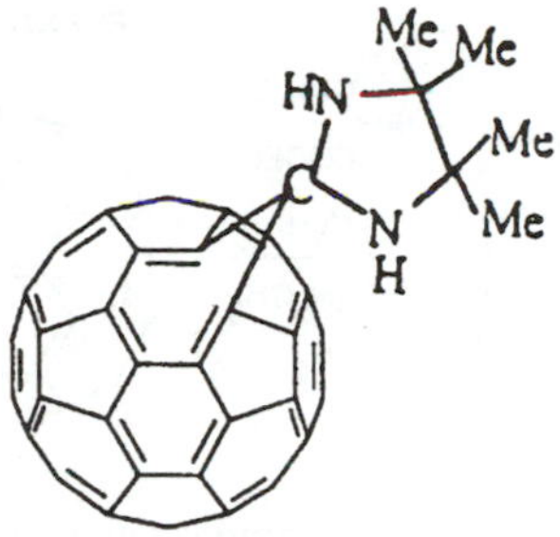

9.21

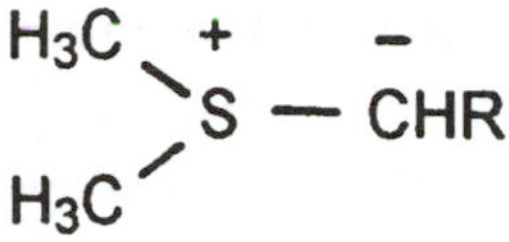

9.22

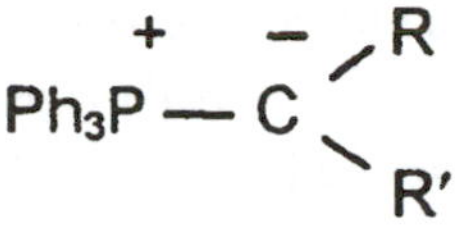

9.23

4,4,5,5-tetramethylimidazolidine-2-thione and DL-valine. This is believed to produce the carbene **9.20** which inserts in to the 5,6-bond to give **9.21**.[52]

9.1.1.2 *Formation of Methanofullerenes Through the Use of Ylides*

The negatively charged carbanionic centres in ylides **9.22, 9.23,** readily react with [60]fullerene to give good yields of methano[60]fullerenes after respective loss of dimethyl sulphide and triphenylphosphine.[53] The former has been used to add an asymmetrically-substituted methano group across the 1,2-bond of [70]fullerene, resulting in chiral products, resolved chromatographically (for R = CONEt$_2$) using a chiral Whelk-O1 HPLC column.[54] This is a resolution technique that is being employed increasingly in fullerene chemistry (see Secs. 9.1.1.3 and 9.1.1.5).

9.1.1.3 *Formation of Methanofullerenes by α-Halocarbanion Addition*

Whilst the reactions described in Sec. 9.1.1.1 have increased our understanding of the chemical behaviour of fullerenes, interest has focused particularly on

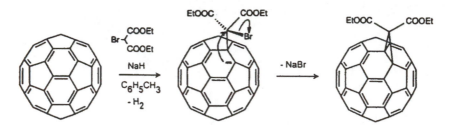

Fig. 9.4 Cyclopropanation of fullerenes by α-halocarbanions.

the use of cycloaddition of α-halocarbanions produced by proton removal from precursors, e.g. by NaH (the Bingel reaction).[55] This (the first step of which is effectively a Michael addition) has the advantage of producing only methanofullerenes and the method, which involves an intramolecular nucleophilic substitution of the side chain by the intermediate fullerene anion (Fig. 9.4), can be adapted to produce a wide variety of derivatives using easily-obtained starting materials, examples being diethyl bromomalonate, methyl 2-chloroacetylacetate, bromomethyl phenyl ketone, α-chlorobenzyl phenyl ketone,[55] and ethyl bromocyanoacetate.[26] Diazabicyclo[5,4,0]undec-7-ene (DBU) is also used as a base, and the addition of CBr_4 can improve improved yields when long side chains are present.[56] In the presence of iodine and DBU, the unhalogenated malonic acid mono-esters will also undergo reaction, giving addition of $ClCO_2R$, though the mechanism may not be straightforward here.[57] Another variation uses nitromethane as the carbanion precursor, proton removal being effected by triethylamine.[26]

These additions tend to be more selective than some others and so may display atypical regiochemistry. For example, monoaddition to [70]fullerene takes place exclusively across the 1,2-bond;[55] the reaction has also been used to characterise pure mono-adducts of (chiral) [76]fullerene. Use of an optically active bromomalonate derivative (and separation with a chiral HPLC column) leads to the isolation of three isomeric pairs of diastereoisomeric mono-adducts and one other mono-adduct. Five have C_1 symmetry and probably involve addition across either the 4,5-, 14,15-, or 12,13-bonds and two others have C_2 symmetry resulting from addition across the 1,6-bond.[58] All of these bonds are predicted to have high π-densities,[59] but the extent to which local curvature

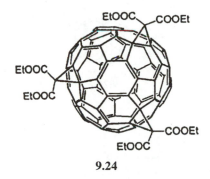

9.24

affects the results is unclear because different regiochemistry is observed in other cycloadditions of [76]fullerene (Sec. 9.4.1). The reaction has also been used to prepare the first characterised derivatives of [78]fullerene (bis- and tris-adducts of C_v-(I)- and D_3 isomers).[58] Addition again occurs across the positions of predicted high π-densities,[59] for example across the 7,8-bond and its two equivalents in the C_3 symmetry derivative **9.24** of the D_3 isomer.

Reactions with α-halocarbanions have been used extensively to study other poly- and tether-directed additions (Sec. 9.1.1.5).

9.1.1.4 *Miscellaneous Reactions Which Produce Methanofullerenes*

Reaction of the Fischer carbene complex $Me(OMe)C = Cr(CO)_5$ with [60]fullerene produces the 1,2-methano[60]fullerene with methyl and methoxy groups substituted at the methylene group. The yield is low and only one example is known.[60] A method which potentially has wider applicability involves reaction of the fullerene with the silyl ether, $4\text{-}MeOC_6H_4C(OSiMe_3) = CH_2$, in the presence of KF/18-crown-6, to give the methanofullerene with COAr substituted at the methylene group; oxygen appears to be necessary for successful reaction. The reaction is thought to occur by tautomeric formation of a carbanion ($MeOC_6H_4CO.CH_2^-$) following removal of the trimethylsilyl group, which then adds to the cage in the manner described in Sec. 9.1.1.3, with subsequent oxidative removal of hydrogen and cyclopropane ring formation.[61]

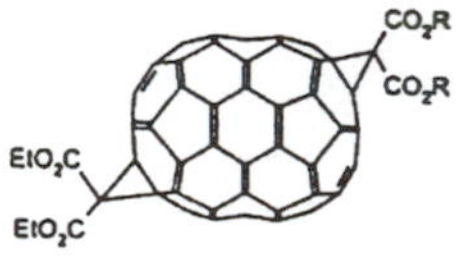

9.25

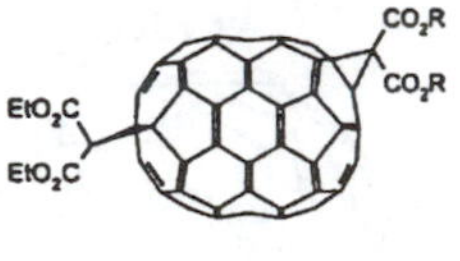

9.26

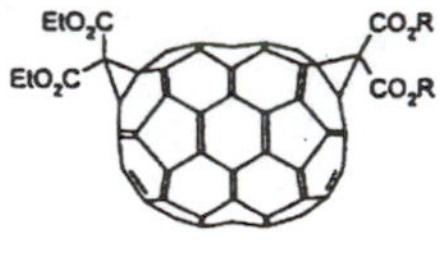

9.27

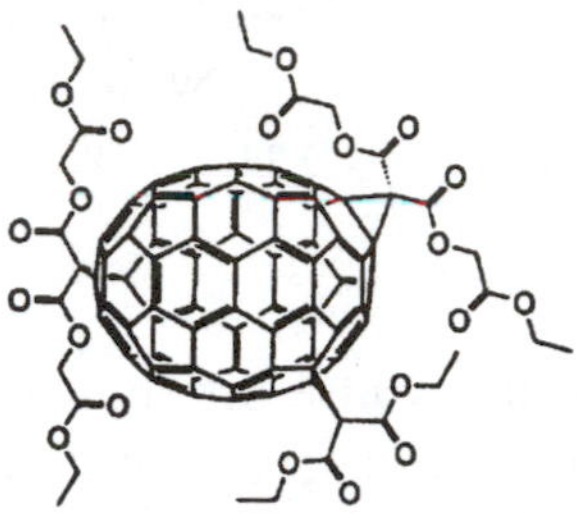

9.28

9.1.1.5 *Polyaddition and Tether-Directed Reactions*

In view of the selective addition of bromomalonate ion to the 1,2-bond in [70]fullerene, bis-cyclopropanation should, and indeed does give the three isomers **9.25–9.27**, which can arise from reaction across the 1,2-bond at one end of the molecule and each of the three equivalent (*a*) bonds at the other.[55] The relative yields, respectively 13:68:19, show the main product to have the same addition pattern (across the 1,2- and 56,57-bonds) as in a bis-iridium complex (see Sec. 11.2). This suggests the involvement of an as-yet unidentified electronic factor.

   A similar diadduct product ratio is obtained in reaction of [70]fullerene with 2-bromopropandioate addends,[62] the 1,2;67,68 derivative being the minor isomer, and the chiral 1,2;56,57- and 1,2;67,68-diadducts (chirality of [70]fullerene diadducts had been predicted by the writer),[63] being resolved through the use of chiral malonates as addends. Addition to the tris- and tetra-adduct level using the 1,2;41,58- and 1,2;67,68 diadducts as precursors, and diethylmalonate as addend, results in the addends in each case adding across a *c,c*-bond at each end of the molecule and in an equatorial position relative to the addend already located across an *a,b*-bond; **9.28** is the trisadduct resulting from addition to the 1,2,67,68-diadduct.

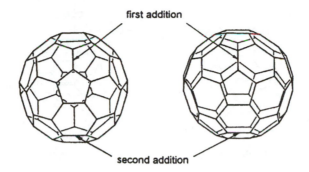

Fig. 9.5 The equatorial regioselectivity found for diaddition of bulky addends at the poles of [60]- and [70]fullerenes.

This is particularly interesting because it is sterically driven (in the absence of steric hindrance, further addition would occur across an adjacent *a,b*-bond, as in hydrogenation, Sec. 4.2.2). Consequently it parallels exactly the equatorial di-addition found with [60]fullerene when bulky addends are involved, Fig. 9.5).

The effect of steric hindrance is shown for example in polyaddition to [60]fullerene. Seven of the eight bis-adducts that can be obtained by addition of $C(CO_2Et)_2$ across a 6,6-bond (see Fig. 2.10) of [60]fullerene have been isolated and characterised.[64] Steric hindrance prevents the formation of the *cis*-1 product, and may account for the low yields of the two other *cis* products. The *e* isomer (which notably is on the pathway to the octahedral hexa-adduct) is formed in highest yield, the *trans*-3 being the second most abundant isomer. Note that the equatorial isomer is statistically twice as likely to be formed as any other, whilst the *trans*-1 product, which is statistically *disadvantaged*, is formed in low yield; this latter is also a feature of the formation of bis-iridium complexes (Sec. 11.2). The coefficients of the frontier orbitals indicate that addition at the *cis*-1, *e* and *trans*-3 positions should be favoured.[65]

Tris addition produces two chiral products (**9.29** and **9.30**) derived by further addition to the two main bis-adducts,[64] and these are described as *e,e,e* and *trans*-3, *trans*-3, *trans*-3, respectively. The former is further along the pathway to the octahedral hexa-adduct, whilst **9.30** is particularly interesting because it has the same regiochemistry found for the only isolated (and unexpected) isomer of $C_{60}H_6$ (Sec. 4.2.3). This may indicate that the catalytic conditions involved

9.29   9.30

in the formation of the latter gives rise to steric hindrance. These enantiomeric [60]fullerene derivatives can be separated using the chiral HPLC technique (see Sec. 9.1.1.2);[66] the corresponding malonic acid derivatives, formed by hydrogenolysis of the esters, have high water solubility.[67]

Control of the sites of polyaddition are an ongoing problem in fullerene chemistry, and two methods have been developed insofar as the Bingel reaction and variants is concerned. One is to pre-add substituted anthracenes to the cage (anthracene adds very readily and the extra groups increase the reactivity and the steric bulk), and then carry out the reaction with bromomalonate etc. The anthracene group can be removed subsequently (in a retro Diels-Alder reaction), by heating at elevated temperatures. In one example, the resulting mixture of diadducts after removal of 2,6-dimethoxyanthracene has a lower *trans* content and higher *cis* content compared to the reaction carried out without the blocking group.[68] Another involves the use of 1,9-dimethylanthracene, which directs addends to the equatorial positions; removal of the directing addend and one final reaction with bromomalonate, produces the hexakis malonate adduct, **9.31**.[69]

An even more remarkable use of anthracene as a directing template leads to the D_{2h} tetramalonate derivative, **9.38** described below. Use of the template addition method also permits formation of the pentamalonate adduct, in which various further reactions can give corresponding hexa-adducts in which the addends occupy all of the octahedral sites. The double bond for the final addition is considerably activated towards addition compared to [60]fullerene itself. This lead for example to the formation of the first fulleren-1,2-diol (from acid-cleavage of the intermediate osmate ester — see also Sec. 11.1), and in turn to the first dioxetane derivative.[70]

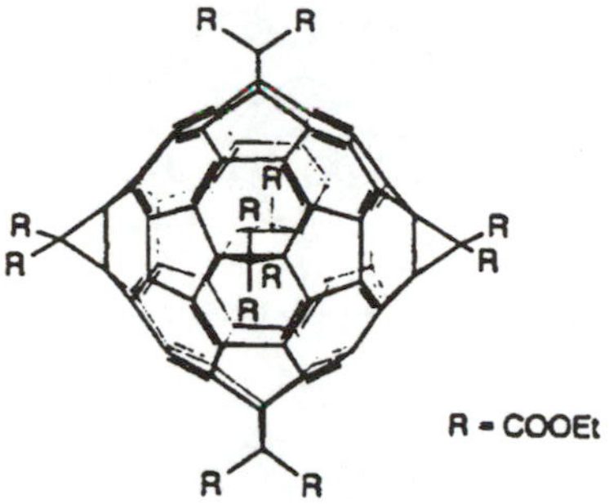

R = COOEt

9.31

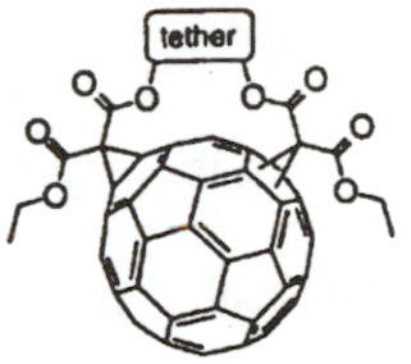

9.32   **9.33** 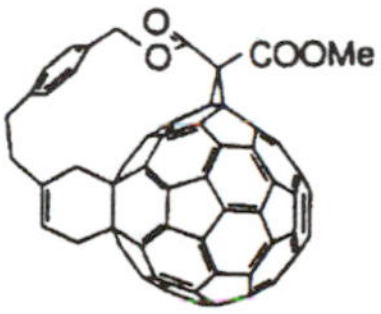   **9.34**

The second method is to use 'tether direction'. In this, addition of a chain takes place at some point in the fullerenes and then a further point in the chain carries out further additions at a location on the cage that is restricted by the length of the tether, e.g. **9.32**. An example is the formation of the *cis*-2 bisadduct **9.33** in up to *ca.* 30% yields, by using a tether such as *o*, *m*, or *p*-xylyl groups.[71] A sequence with wider application involves a Diels-Alder dienophile at the end of a tether of length calculated to produce addition at the equatorial positions, giving structure such as **9.34**. Subsequent additions occur at the equatorial positions giving ultimately the hexakis adduct **9.35**. Strategies developed for removal of the cyclohexene rings are shown in Fig. 9.6 which in this instance gives the D_{2h} symmetrical tetrakis(methano)[60]fullerene **9.38**, a stereochemical arrangement that had not been achieved previously. The process involves oxygen addition to **9.35** to form the allylic hydroperoxides (**9.36**, R = OOH), deoxygenation to give the corresponding alcohols, acid-catalysed dehydration to give the bis (cyclohexa-1,3-diene, **9.37**). Finally, a retro Diels-Alder reaction and transesterification with EtOH/THF/K_2CO_3 produced **9.38**.[72]

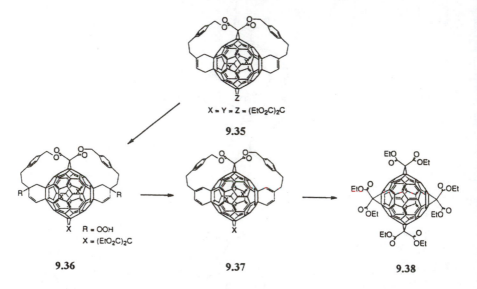

Fig. 9.6 A procedure for removal of directing/blocking groups (see text).

This compound can now be more easily prepared formed by making use of the bis anthracene derivative (which forms spontaneously from the monoadduct by disproportionation) having anthracene addends in an antipodal relationship. Following the tetramalonate addition, the anthracenes are easily removed just by heating in the absence of oxygen.[73]

Through the use of the Bingel reaction, with the aid in some cases of the tether directing procedure, a range of complex [60]fullerene structures have been made. Some examples which demonstrate the variety and complexity are shown in **9.39–9.43**.[74]

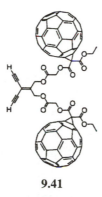

9.41

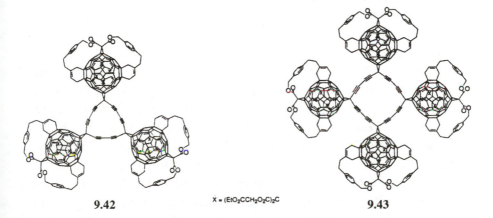

9.42   X = (EtO₂CCH₂O₂C)₂C   **9.43**

Electrochemical investigations of mono- to hexa-adducts formed *via* the tether directed method show that with increasing number of addends, reduction becomes more difficult (and less reversible) and oxidation correspondingly easier the higher the addition level,[72,75] which follows from the increasing electron supply to the cage with increasing number of addends.

9.1.2 *Addition of Nitrogen: Formation of Azahomofullerenes and Epiminofullerenes*

Nitrogen may be either added across a 6,6 π-bond or inserted into a 6,5 σ-bond. With [60]fullerene, the former gives 1,2-epimino[60]fullerene (**9.44,**

9.44    9.45

otherwise known as an aziridinofullerene), whilst the latter gives 1a-aza-1(6)a-homo[60]fullerene (**9.45**, R = H otherwise known as an azafulleroid). The insertion reaction allows greater bond angles at nitrogen which therefore tends to be sp^2-hybridised in azahomofullerenes (and sp^3-hybridised in epiminofullerenes).

Two main methods are used for introducing nitrogen to the cage: reaction with azides (which proceeds via an initial [3 + 2] cycloaddition followed by nitrogen extrusion) gives mainly the azahomofullerene,[76,77] whilst reaction either with nitrenes (formed by either thermal[78,79] orphotochemical[80] decomposition of azidoformates, by lead tetra-acetate oxidation of N-aminophthalimide[81] or by α-elimination of O-4-nitrophenylsulphonylalkyl-hydroxamic acids,[82] gives mainly the epiminofullerene. For example heating [60]fullerene with a series of azides RCH_2N_3 gives the azahomo[60]fullerenes **9.46**,[76] (which are, as expected, more electronegative than their carbon analogues).[76,83] Reaction of [60]fullerene with methyl glycylglycinate in the presence of bromine and tetrabutylammonium bromide produces a homoazafullerene dipeptide C_{60} > $NCH_2CONHCH_2CO_2Me$; the reaction is thought to proceed through oxidation of the terminal amino group of the ester into an nitrene.[84]

The parent 1,2-epimino[60]fullerene (aziridino[60]fullerene), **9.44** (R = H), obtained by elimination of isobutene and CO_2 from **9.44** (R = CO_2t-Bu),[80,85] can be used as a nucleophile in reactions with acyl halides and anhydrides to give e.g. **9.47**, so providing a useful route to further fullerene functionalisation.[78,86] Carbamate esters (**9.44**, R = CO_2Et or $CO_2$2,4,6-tri-t-butylphenyl) can also be obtained by direct reaction of the fullerene with acylnitrenes (from the azide precursors), and on heating, ring expansion of the

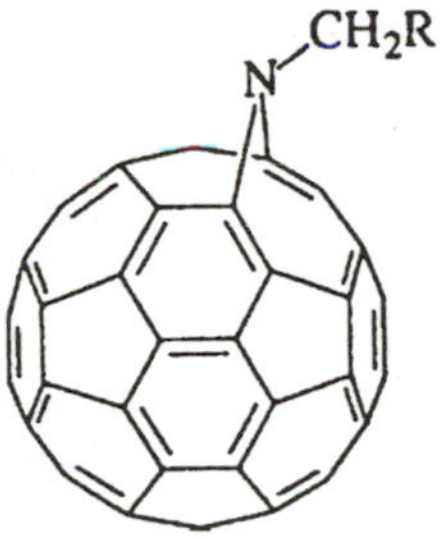

9.46 [R = O(CH₂)₂SiMe₃; Ph; 4-MeOC₆H₄; 4-BrC₆H₄].

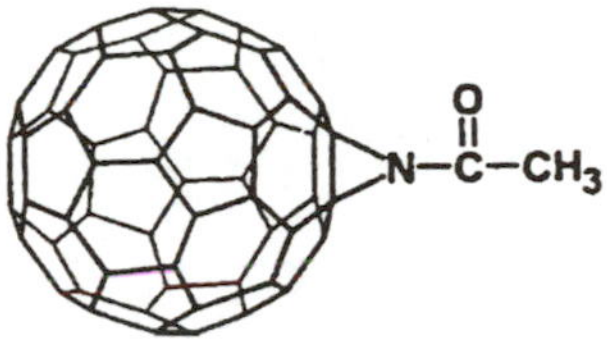

9.47 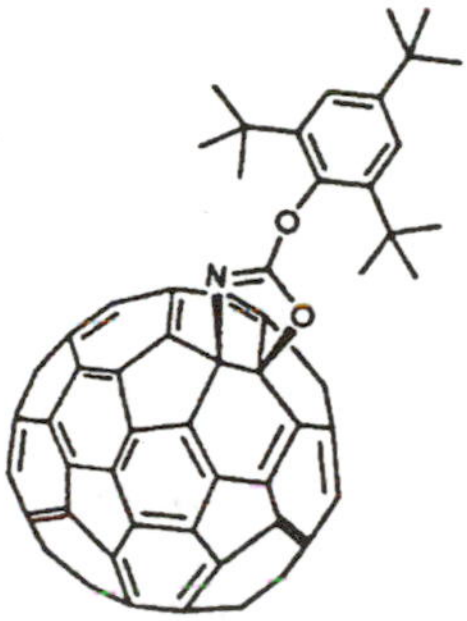   9.48 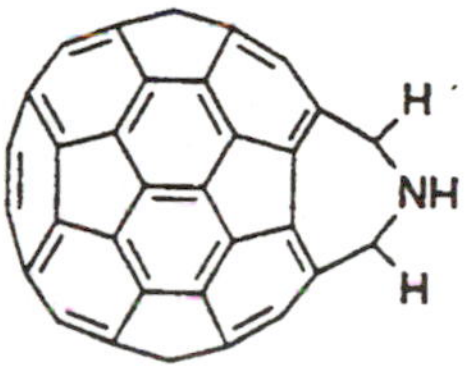   9.49

aryl derivative occurs to give the corresponding 2-(2,4,6-tri-t-butylphenoxy)-oxazolo[4,5:1,2][60]fullerenes **9.48**.[2,78] Aziridine ring expansion also occurs on reduction with Zn/HOAc of carbamate esters **9.44** (R = OCOtBu) to give **9.49** (after removal of the ester group by reaction with TFA), though this rapidly and spontaneously oxidises back to **9.44** (R = H).[87]

Azahomo[70]fullerenes can be prepared by reacting [70]fullerene with substituted methyl azides.[88] Addition of the triazoline moiety takes place across the 1,2-bond (each way) and across the 5,6-bond. Subsequent thermolysis produces from each, two azahomo[70]fullerenes (**9.50**, **9.51**), and two epimino[70]fullerenes (**9.52**, **9.53**). Note that **9.51** is obtained from initial azide addition to either of the 1,2- or 5,6-bonds.

Various azahomo- and epimino[60]fullerenes with aromatic and heteroaromatic substituents attached to nitrogen are known.[2,83,89,90] One azahomofullerene, with a 9-(4-benzyl)acridine group attached (via methylene)

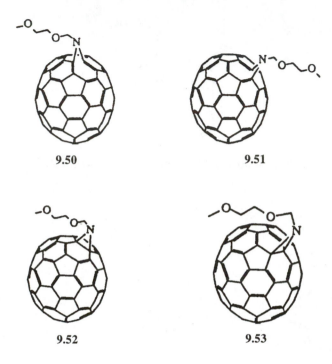

9.50   9.51

9.52   9.53

to the nitrogen, has enhanced DNA-cleaving ability (compared to [60]fullerene).[89] Another having a 4-succinimidoxycarbonylphenyl group attached to the nitrogen shows a quasi-reversible oxidation during electrochemistry, the first example of such reversibility for a fullerene.[83]

A particularly interesting development in this field of fullerene chemistry is the cage opening that becomes possible with the azahomofullerenes.[91] Just as those alkenes, which possess electron-supplying oxygen adjacent to the double bond (vinyl ethers), readily undergo oxidative cleavage *via* an intermediate 1,2-dioxetane, to give two carbonyl groups,[92] so the presence of adjacent (and even more electron-supplying) nitrogen produces similar cleavage. This happens with the azahomofullerenes, resulting in an eleven-membered hole in the cage (Fig. 9.7), just as is found with corresponding oxygen compounds (Sec. 8.6.2); the presence of the substituent on nitrogen makes the hole slightly more restricted than for the oxygen analogue. The cage-opened derivative is chiral and so can be resolved into its enantiomers using a chiral HPLC column.[93]

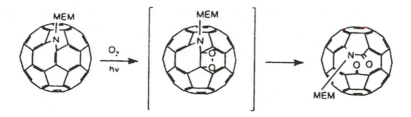

Fig. 9.7 Oxidative formation of cage-opened [60]fullerene from *N*-methoxyethoxy-methylazahomo[60]fullerene.

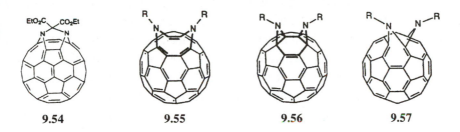

| 9.54 | 9.55 | 9.56 | 9.57 |

Three types of bis azahomofullerenes are now known:

(i) Those with connected nitrogens, e.g. **9.54**, are prepared by heating bis azo compounds $N_3(CH_2)_nN_3$, $n = 1-3$ (some with substituents on carbon), to *ca.* 135°C with [60]fullerene.[94] Here, the nitrogens insert into 6,5-bonds in a given pentagon.

(ii) By contrast, derivatives with unconnected nitrogens are obtained by more gentle heating of the fullerene with two equivalents of e.g. a monoazidoformate. However, because of repulsion by non-bonding orbitals, the nitrogens *insert* across 6,6-bonds to give structures **9.55** (R = CO_2Et, CO_2t-Bu).[95] This places the nitrogens further apart than if they were inserted into 6,5-bonds, i.e. both were in a five-membered ring. A related argument concerning repulsion between non-bonding orbitals of oxygen, accounts for the preferred structure for $C_{120}O_2$.[96] An interesting aspect is that derivatives **9.55** (R = H) undergo an intra-ring Diels-Alder reaction to give the valence isomer **9.56**.[95]

(iii) Vigorous heating (12 h reflux in chlorobenzene) of [60]fullerene with two equivalents of *C*-substituted methylazides produces epiminofullerenes and azahomofullerenes along with the other type of bis adduct, **9.57**, in which nitrogen has here inserted into two adjacent 6,5-bonds.[97] The corresponding double insertion occurs in the pentagonal cap of [70]fullerene.[98] The importance of compounds **9.57** and homologues is that they provide an easy access to heterofullerenes (Chap. 13).

9.1.3   *Addition of Oxygen: Formation of Epoxides*

Formation of epoxides is an area of fullerene chemistry that is rapidly growing in importance, and is likely to continue to do so, given that fullerenes readily form epoxides, even on standing in air. They are also formed during the arc-discharge process for making fullerenes, due to the presence of traces of oxygen in the reactors, and can be separated from the soot extract by HPLC.[99,100] Thus far there has been almost no chemistry carried out with oxides, for two main reasons. First, the compounds tend to degrade on standing, though no characterisation of the products (apart from isolation of C_{119} produced by high temperature degradation, Sec. 12.5) has yet been made. Solutions in toluene deposit material on the flask walls at room temperature, and this is soluble only in 1,2-dichlorobenzene.[101] Secondly, controlled formation of the mono-epoxide is not yet possible, so that time-consuming HPLC separations are necessary. As successive oxygen atoms are added to [60]fullerene, the colours of toluene solutions change from magenta through pink, yellow, orange to brown.[101]

There are a wide range of methods of formation of fullerene epoxides. These comprise photooxidation of fullerene in the presence of photosensitizers,[102–104] or by using chemically-generated singlet oxygen,[105] oxidation by either dimethyldioxirane,[106] iodosobenzene in the presence of metal catalysts,[107] or by methyltrioxorhenium-hydrogen peroxide,[108] electrochemical oxidation,[109] ozonolysis,[110–113] and oxidation with *m*-chloroperbenzoic acid.[114,115] The last method is by far the easiest to carry out, and is the only one to be considered seriously for chemical studies.

Like most other fullerene reactions, oxidation does not go to completion (but also gives polyaddition), and so separation from unreacted fullerene (and higher oxides) is necessary, being most easily accomplished by HPLC. Prepurification by column chromatography using alumina is not recommended

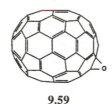

9.58   **9.59**   **9.60**

because it causes reversion to the parent fullerene,[102] though silica gel is apparently satisfactory.[113]

The structures of 1,2-epoxy[60]fullerene (**9.58**),[102] and of 1,2- and 5,6-epoxy[70]fullerenes (**9.59**, **9.60**)[99,104] are as shown, the isomer ratio of the latter pair being *ca.* 43:57 and apparently independent of the method of formation; the greater reactivity of the 1,2-bond compared to that of the 5,6-bond is consistent with all other reactions of [70]fullerene. Likewise the greater reactivity of [60]fullerene compared to [70]fullerene towards oxygen parallels the relative reactivities of these two molecules in many reactions. The higher fullerenes have been reacted with ozone, but the products have not been characterised.[110]

The number of oxygens that can be added to [60]fullerene varies somewhat with the oxidation method, but up to five can be readily detected,[4,107,109–111] and two to [70]fullerene.[4,110] However MALDI-TOF analysis of the products of the reaction with *m*-chloroperbenzoic acid, shows the addition of eleven oxygens to [60]fullerene and five with [70]fullerene,[116] the latter result being confirmed in the reaction with ozone.[112] In the formation of epoxides from fluoro[60]fullerenes, up to eighteen oxygens are present, and *ca.* eleven from fluoro[70]fullerenes;[117] eighteen oxygens are also added in the gas-phase reaction between [60]fullerene anions and ozone.[118]

The structure of $C_{60}O_2$ has been determined as 1,2;3,4-diepoxy[60]fullerene **9.61**, this being the expected isomer for the same underlying reason (see Sec. 3.1) that gives rise to 1,2,3,4-tetrahydro[60]fullerene (Sec. 4.2.2) as the main isomer on reduction of [60]fullerene.[115] Likewise $C_{60}O_3$ consists of a mixture of 1,2;3,4;9,10-triepoxy[60]fullerene **9.62**, and 1,2;3,4;11,12-triepoxy[60]fullerene **9.63**, in *ca.* 3:2 ratio.[107] Note that respectively, these have the 'S' and 'T' patterns expected as a result of bond localisation (Sec. 3.1) and

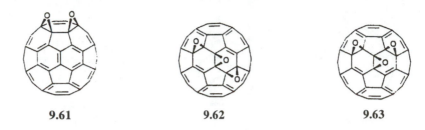

9.61   **9.62**   **9.63**

which underlie the formation of $C_{60}H_{18}$ (Sec. 4.2.4) and $C_{60}F_{18}$ (Sec. 7.1.1.2). The possibility exists therefore that $C_{60}O_9$ and $C_{60}O_{18}$ are each isostructural with the $C_{60}X_{18}$ and $C_{60}X_{36}$, respectively.

Epoxides are also formed spontaneously when reactions are performed with halogenofullerenes. For example they are present in many arylated derivatives produced by Friedel-Crafts reactions of bromo- and chlorofullerenes with aromatics (e.g. Sec. 8.6.4). They arise from nucleophilic substitution of halogen by OH and subsequent elimination of Hhal (see Fig. 7.1). This is particularly evident with fluorofullerenes which are the most reactive halogenofullerenes towards nucleophilic substitution, and the subsequent elimination of HF is evident as etching of containers if the materials are not stored under anhydrous conditions.

Studies of the reactivity of the epoxides is in its infancy. The ability to produce singlet oxygen diminishes rapidly with increase in the number of oxygens on the cage.[119] The oxygen can be removed by reacting with triphenylphosphine,[120] and photodegradation is faster than for [60]fullerene itself.[110] Heating a mixture of $C_{60}O_{1-3}$ and C_{60} under various conditions, produces both $C_{120}O$ and $C_{120}O_2$ (see Sec. 12.5) and also $C_{180}O_2$.[121] Heating $C_{60}O$ alone produces CO_2 and C_{119} in what appears to be a bimolecular reaction,[101] and C_{119}, C_{129} and C_{139} are formed by heating various fullerene/fullerene oxide combinations.[103]

9.1.4 *Addition of Silicon: Formation of Fullerene Siliranes*

Silamethano[60]fullerenes (silirano[60]fullerenes) and also bis- and tris adducts are obtained by reaction of [60]fullerene with photochemically generated bis(2,6-di-isopropylphenyl)silylene (Fig. 9.8).[122]

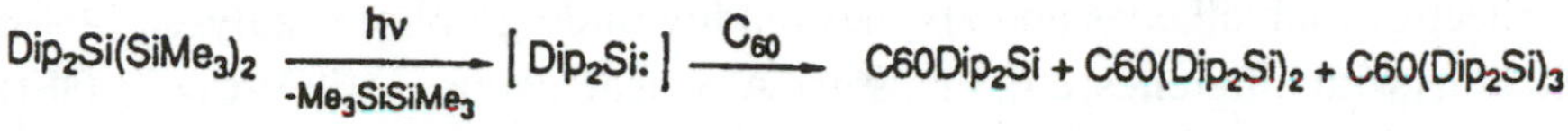

$$Dip_2Si(SiMe_3)_2 \xrightarrow[-Me_3SiSiMe_3]{h\nu} \left[Dip_2Si: \right] \xrightarrow{C_{60}} C_{60}Dip_2Si + C_{60}(Dip_2Si)_2 + C_{60}(Dip_2Si)_3$$

Fig. 9.8 Route for formation of a silamethano[60]fullerene.

9.2 [2 + 2] Cycloadditions

Photochemical [2 + 2] cycloadditions take place with fullerenes, giving products with four-membered rings (in some cases containing a double bond) fused to the fullerene cage. Further ring-opening reactions may occur due to the strain in these rings. For orbital symmetry to be conserved, suprafacial addition should be observed if the reactions are fully synchronous; when this is not the case, biradial intermediates are implicated.

9.2.1 *Additions by Alkenes*

In the photocatalysed reaction between [60]fullerene and cyclic enones (which gives products possessing up to seven addends), both *cis* and *trans* monoaddition products are obtained, indicating that the intermediate is a biradical (Fig. 9.9).[123]

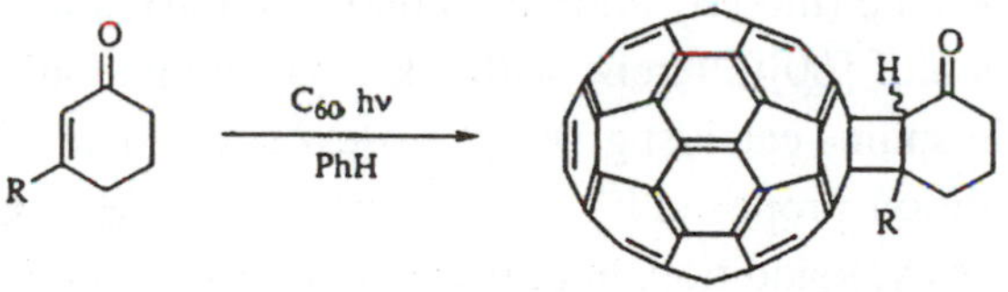

Fig. 9.9 Formation of *cis* and *trans* products from reaction of cyclic enones with [60]fullerene.

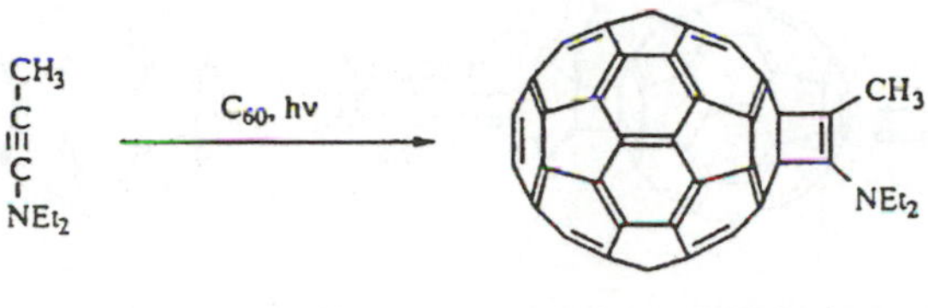

Fig. 9.10 Photocatalysed reaction of [60]fullerene with electron-rich dienes.

Electron-rich alkenes and alkynes readily undergo photocatalysed [2 + 2] reactions with fullerenes, e.g. Fig. 9.10. A typical alkene is $(EtO)_2C = C(OEt)_2$ and diaddition of this with [60]fullerene takes place at the equatorial position, typical for a bulky addend (see Sec. 3.2). Monoaddition with [70]fullerene occurs preferentially across the 1,2-bond, the 1,2-:5,6-reactivity ratio being estimated here to be a very high (28); hydrolysis of the diketal products could not be achieved (*cf.* **9.11**, Sec. 9.1.1.1).[124] Other electron-rich addends employed in reaction with both [60]- and [70]fullerene are $Et_2NC \equiv CNEt_2$ and $Et_2NC \equiv CSEt$. With [70]fullerene, diadducts are obtained, two of which, produced in *ca.* 9:1 ratio, have C_2 symmetry. By analogy with other reactions giving di-addition to [70]fullerene (Secs. 9.1.1.5 and 11.2) the predominent isomer is assumed to be that resulting from reaction across the 1,2- and 56,57-bonds. Oxidative cleavage of the double bond in the resultant four-membered ring produces diamides, which in turn can be converted to anhydrides by reaction with *p*-toluenesulphonic acid. The anhydrides are unstable in the presence of water due, probably, to hydrolysis to the diacids which rapidly decarboxylate owing to the strong electron withdrawal by the cage, giving the dihydrofullerenes, which are themselves readily oxidised.[125]

Addition of dimethyleneketene acetals gives products which are readily ring-opened by acid (Fig. 9.11),[126] whilst addition of deuterium labelled 4-methoxystyrene **9.64**, gives both *cis* and *trans* products showing that a diradical intermediate (thereby allowing bond rotation), must be involved.[127] However, reaction of [60]fullerene with alk-2-ynoate and alk-2-ynone in the presence of a phosphine catalyst gives [2 + 3]cycloaddition,[128] rather than the [2 + 2] cycloaddition proposed.[129] Allyl compounds $CH_2 = CH.CH_2Ar$ also do not take part in cycloaddition, but instead undergo the 'ene' reaction giving

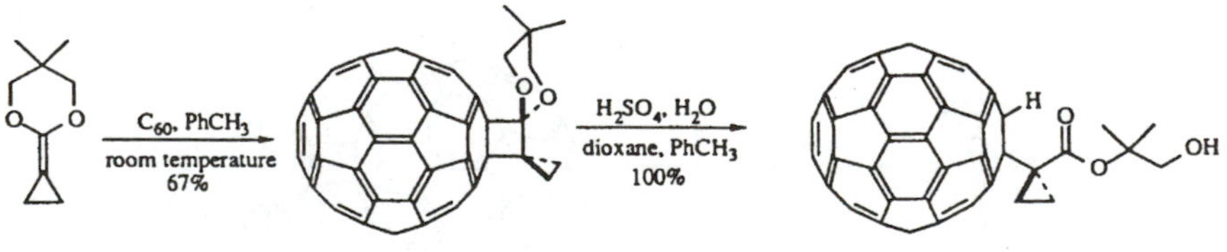

Fig. 9.11 Formation of esters via reaction with dimethylene ketene acetals and acid.

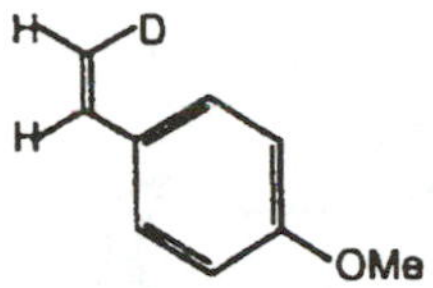

9.64

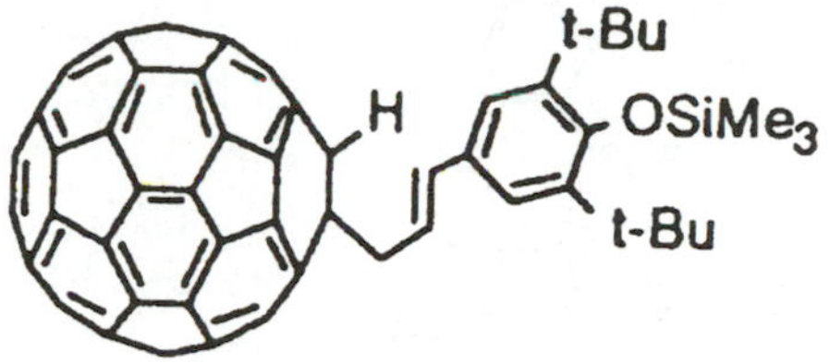

9.65

e.g. **9.65**.[130] A silyl ketene acetal $Me_2C = C(OMe)OSiMe_3$ also does not undergo [2 + 2] photoaddition, due probably to steric hindrance, but instead (in the presence of moisture) gives 1,2-addition of H and $C(Me)_2CO_2Me$ following loss of trimethylsilanol ($\rightarrow$ silicone), so providing a useful route to fullerene-substituted esters.[131]

9.2.2 *Additions Involving Pseudo Alkenes (4-Membered Rings etc.)*

The strain in three- and four-membered rings causes them to readily undergo additions in general, and this is manifested also in their reactions with fullerenes. Examples are the reaction of benzocyclobutenols,[132] 1,2-dicarbomethoxycyclobutadiene (from an iron tricarbonyl precursor),[133] and cyclotetra-silanes and -germanes,[134] giving e.g. **9.66**, **9.67** (E = CO_2Et) and **9.68**, respectively. Reaction with quadricyclane, which contains two strained three-membered rings, gives **9.69** and notably here, the double bond in the addend and which is not sterically hindered by the cage, undergoes *trans* electrophilic addition, in contrast to that which results from the addition of cyclopentadiene (see Sec. 9.4.5).[135]

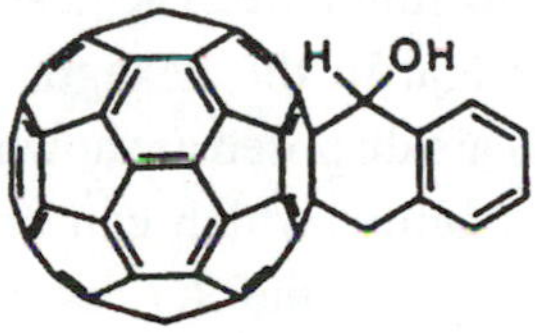

9.66

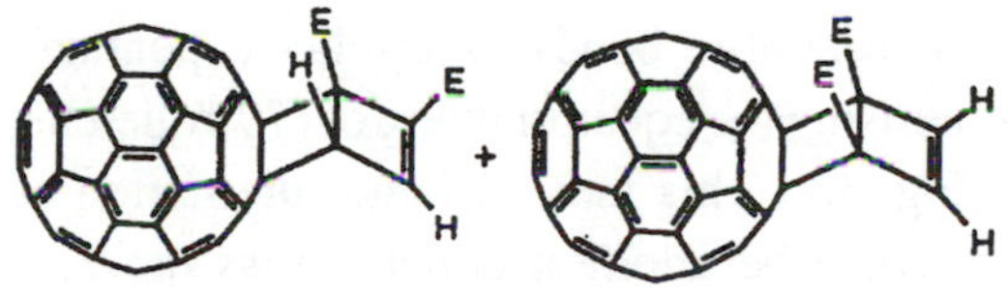

9.67

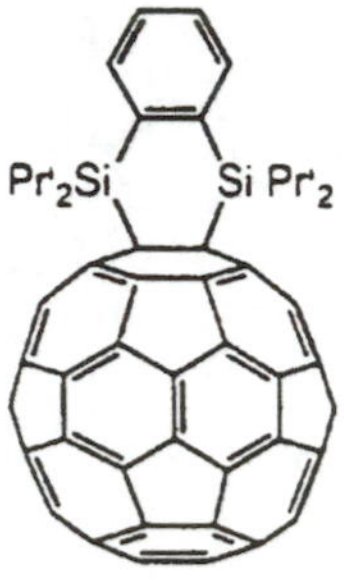

9.68

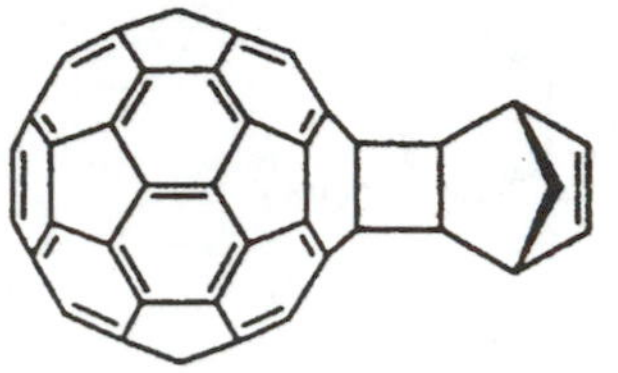

9.69

9.2.3 *Addition of Benzyne*

Addition of benzyne shows behaviour typical of many reactions with fullerene, namely that little or no reaction occurs when approximately equivalent amounts of reagent are used. However, on increasing the quantity of non-fullerene reagent, reaction occurs rapidly resulting in polyaddition. Such is the case on reaction of fullerenes with benzyne.

With [60]fullerene up to six benzynes add to the cage,[136,137] and the monoadduct has been isolated and characterised.[136] Although the structure of the hexa-adduct has not been determined, it is reasonably certain that the addends are located at the octahedral sites.

With [70]fullerene, up to ten benzynes add to the cage.[138] The monoadduct consists of four isomers, the first and so far only example of four mono-addition products with this fullerene. Structural characterisation by ^{1}H NMR indicated that addition occurred across the 1,2-, 5,6, 7,21 and 7,23 positions, the latter being a homologue of triptycene. However, more recent work involving [1]C NMR analysis[139] shows that the supposed 7,23-isomer is in fact the 7,8-isomer (both give the same ^{1}H NMR pattern, having C_s and C_2 symmetries, respectively), but different ^{13}C NMR patterns. An interesting aspect arises here and is of general applicability to the reactions of [70]fullerene. The need to have double bonds exocyclic to pentagons, means that in the hexagons comprising the equatorial ring of [70]fullerene, the bonds are disposed as shown in Fig. 3.2. This lack of bond alternation confers instability, which can be overcome be addition either across the 7,23-bond (Figs. 3.3 and 7.7, as in chlorination, Sec. 7.1.2.), or across the 7,8-bond (Fig. 3.4) as here. The four products of addition of benzyne to [70]fullerene are thus **9.70–9.73**.

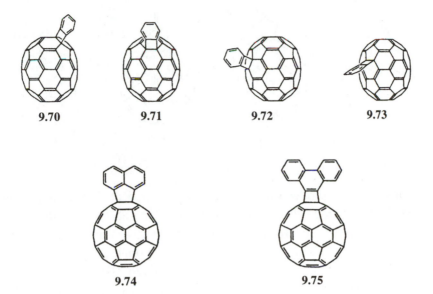

9.70   9.71   9.72   9.73

9.74   9.75

An unexplained result arises in the reaction of 1,8-naphthyne with [60]fullerene which gives **9.74**, the main product having four addends attached to the cage (as in the case of benzyne addition).[140] However up to *ten* of these bulky addends are also added, providing a fascinating problem as to how they can be accommodated on the cage. 9,10-Phenanthryne also adds to [60]fullerene to give **9.75**.[140]

9.3 [3 + 2] Cycloadditions

Addition of X-Y-Z to the cage involves in principle, all possible combinations of C, O, N, S, and Si, many of which have been realised to date. All reactions, except when carbon only is involved, lead to heterocyclic derivatives, many of which may ultimately have important applications.

9.3.1 *Formation of Cyclopentafullerenes*

These compounds can be produced both with and without a double bond in the five-membered ring, **9.76**, **9.77**, respectively. They are for example formed

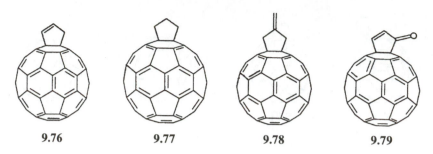

9.76   **9.77**   **9.78**   **9.79**

spontaneously as byproducts in many reactions, as indicated by the appearance of a peak at 762 amu (always accompanied by one at 760 amu due to loss of H_2) in the mass spectrum.[141] Similar derivatives are obtained from [70]-, [78]- and [84]fullerenes by reaction of brominated fullerene mixtures with THF, from which trimethylene is abstracted. The alternative tris-methanofullerene structures are ruled out by the above fragmentation pattern,[5] whilst addition across a 6,6-bond rather insertion into a 5,6-bond is favoured energetically.[142]

The adduct of trimethylenemethane and [60]fullerene, **9.78**, can be prepared by reacting $CH_2 = C(CH_2OAc)CH_2SiMe_3$ with an intermediate palladium-C_{60} complex,[143,144] whilst the enone (**9.79**) is obtained by hydrolysis of its ketal precursor (obtained by heating [60]fullerene with cyclopropenone 1,3-propanediyl ketal).[18,144] The latter reagent (which gives rise to a vinyl carbene intermediate and producing [1 + 2] addition if the reaction temperature is too low), also reacts with [70]fullerene to give addition across the 1,2-bond. Reaction conditions are critical because if traces of water are present, the intermediate ketal rearranges to an ester.[145]

Diaddition involving reagents **9.80** demonstrates how geometrical constraints affect regiochemistry in fullerenes. When $n = 3$, C_s symmetry products arising from addition across the 1,2;3,4 and 1,2;7,21-bonds are obtained, but when $n = 4$, only 1,2;3,4 addition is obtained, whilst no addition occurs for $n = 5$, these differences arising from a combination of the chain length and the interaction of the chain hydrogens with the cage surface. When $n = 6$, addition occurs across the 1,2;16,17-bonds, to give the C_2 symmetry product **9.81**; by modification of the structure of the addend reagent, the tether can be cleaved.[146]

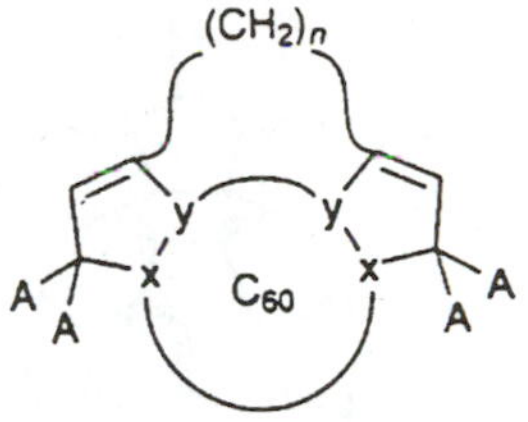

9.80 (A,A = 2,2-dimethylpropan-1,3-diyl acetal)

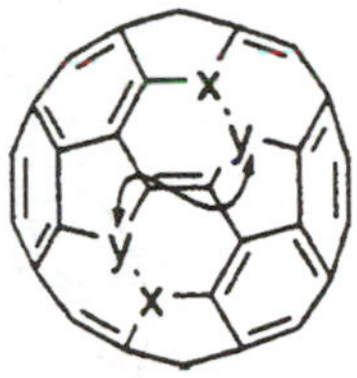

9.81

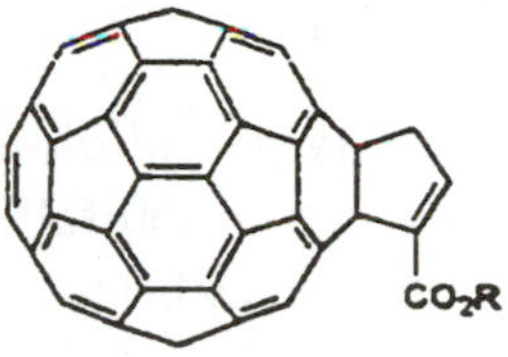

9.82

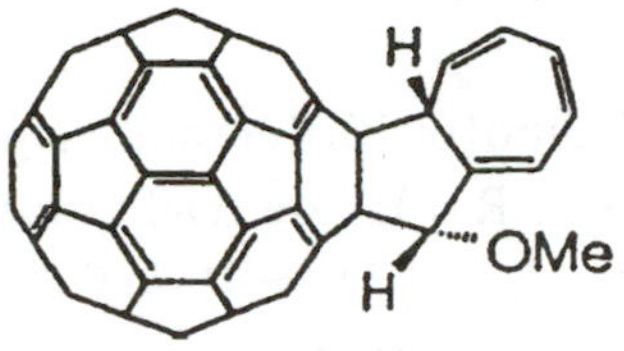

9.83

The cyclopentena[60[fullerene derivative **9.82** (see also Sec. 9.2.1), is formed by [3 + 2] cycloaddition between either buta-2,3-dienoates ($CH_2 = C = CH.CO_2Et$) or but-2-ynoates ($MeC \equiv C.CO_2Et$) and [60]fullerene in the presence of triphenylphosphine.[128] A fullerene-fused cyclopentane derivative (**9.83**, a tetrahydroazuleno[60]fullerene) is obtained from the reaction of 8-methoxyheptafulvene with [60]fullerene, but the mechanism here is actually an [8 + 2] cycloaddition.[147] As expected, electron release from the addend makes reduction by electron addition harder than for the parent fullerene. The addend can be oxidised to give the tropylium ion, and can undergo further addition.

9.3.2 *Formation of Pyrrolidinofullerenes*

The more convenient and steroid-based IUPAC nomenclature is used here and in subsequent sections; *cf.* the Chem. Abstracts system which hitherto has been used by a few authors. These reactions have been the most extensively investigated amongst the [3 + 2] cycloadditions with interest centering on preparation of derivatives having potential electroactive substituents as part of the addend, optical-limiting devices etc.

9.84    9.85

Pyrrolidino[60]fullerenes (e.g. **9.84**, *N*-methylpyrrolidino[3,4:1,2] [60]fullerene, R = H) are readily prepared by reaction of [60]fullerene with azomethine ylides $CH_2 = N^+Me.CHR^-$ [readily prepared by heating aldehydes RCHO with *N*-methyl glycine, (sarcosine, $MeNH.CH_2CO_2H$)]. This route enables the preparation of a wide range of derivatives, e.g. with R = ferrocene, pyridine (β-substitution gives a nicotino[60]fullerenes), various ester groups, and tetrathiofulvene (and analogues);[148,149] the latter derivative and modifications of it (which are slightly harder to reduce than [60]fullerene)[150] exhibit charge-transfer properties. Derivatives here include some with potential electroactive properties e.g. in which R is a covalently-linked transition metal complex, $Ru(bipyridyl)_3^{2+}$,[151] or tetraphenylporphyrin (see also Sec. 12.7), (in this latter work dumbbell structures were also obtained, see Sec. 12.6).[152] Compounds with R = o-$HS(CH_2)_4C_6H_4$ self assemble into a monolayer on a Au(III)/mica surface,[153] a property, which with further development, may have useful applications.

Chirality can be introduced into the molecule through appropriate modification of the addend,[154] and water solubility (*ca.* $2–3 \times 10^{-5}$ M) can be obtained by replacing the methyl group attached to nitrogen by $(CH_2CH_2O)_3Me$ giving products with promising biocide activities.[155] Various derivatives with silicon-containing groups attached either to nitrogen or at the adjacent carbons have been prepared with the object of increasing solubility and ability to be covalently attached to a silicon surface, yet retaining the optical limiting properties of the parent fullerene.[156]

A further variation involves use of aminoacids $H_2NCHRCO_2H$ and aldehydes R′CHO, which give the intermediate azomethine ylide

CHR = N$^+$H.CHR$'^-$, and derivatives **9.85**.[157] PhCH(SiMe$_3$)NH$_2$ can also be used in place of the amino acid (the leaving group here is SiMe$_3$H instead of CO$_2$; xs. BBr$_3$ is required for the reaction to occur), and here both *cis* and *trans* products are possible.[158] Derivatives of the type shown in **9.84** and **9.85** (R all carboalkoxy), can apparently be made from photochemical reactions between [60]fullerene and either alkyl glycinates, or alkyl *N*-methyl glycinates, respectively,[159] but the mechanisms are evidently complex since the carbon counts of the addends increase in an unknown way. They can also be made by the thermal reaction between [60]fullerene, aldehydes (R = *n*-C$_5$H$_{11}$ and *n*-C$_{11}$H$_{23}$) and ammonia, though for R = benzyl, 1,2-addition of benzyl and H occurs instead.[160]

Comparison of the regiochemistry of di-addition with those in the [1 + 2] Bingel reaction (Sec. 9.1.1.5) shows[161] the two reactions to give significantly different addition patterns. In particular much less of the *e* isomer and correspondingly more of the *cis*-3 isomer is obtained in the [3 + 2] cycloaddition. Quite different addition patterns are found also in the reaction with [70]fullerene, the isomer yields being 1,2-(46%), 5,6-(41%) and 7,21-(13%),[162] whereas the [1 + 2] reaction gives exclusive 1,2-addition (Sec. 9.1.1.3). No explanation based on surface curvature can explain the variations between reaction in additions to [70]fullerene and differences in reactivity and hence selectivity of the attacking reagents is a probable cause.[99]

Compounds with NH groups **9.85**, tend to be less stable than those with NMe groups **9.84**, due probably to the tendency of amino groups to react with fullerenes. However, stability can be increased by acylation of the nitrogen.[148,149,163]

Pyrrolidinofullerenes can also be prepared by reaction with aziridines, as shown in Fig. 9.12,[148,163] and replacing *N*-benzyl with *N*-(CH$_2$)$_3$Si(OEt)$_3$ gives a product which shows promise as a stationary phase for HPLC since it binds to silica.[164]

Pyrrolidinofullerenes can also be obtained by CO$_2$ loss from oxazolidinones **9.86** (R = 4-MeOC$_6$H$_4$Ph$_2$C-). Cleavage of the *N*-substituent with trifluoromethanesulphonic acid gives **9.85** (R,R$'$ = H), and subsequent acylation at nitrogen gives products showing promising applications as Langmuir-Blodgett films.[165]

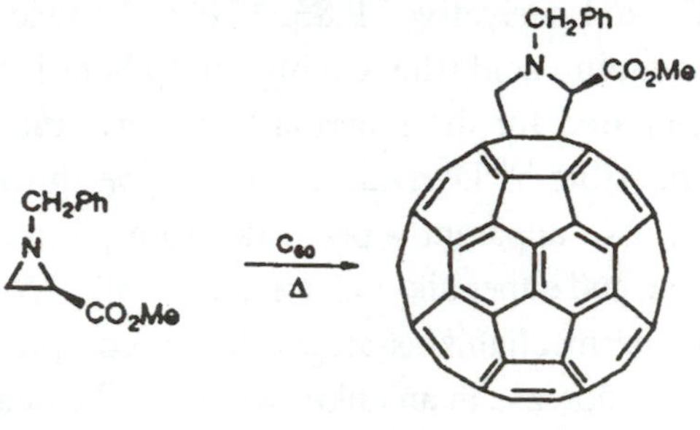

Fig. 9.12 Formation of pyrrolidino[60]fullerenes by ring-opening of aziridines.

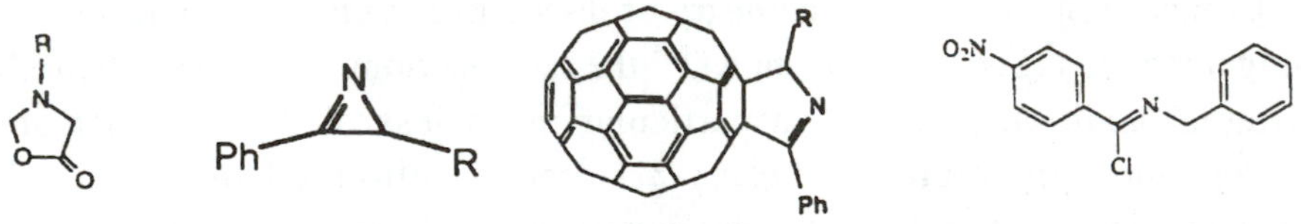

9.86   **9.87**   **9.88**   **9.89**

9.3.3 *Formation of Pyrrolo[60]fullerenes*

Pyrrolo[60]fullerenes have been made in two ways. First by photocatalysed reaction between [60]fullerene and 2,3-diphenyl-2*H*-azirine **9.87**, giving 2,5-dipheny-4,5-dihydro-3*H*-pyrrolo[3,4:1,2][60]fullerene **9.88** (R = Ph).[166] They have also been made by thermal reaction between [60]fullerene and *N*-benzyl-4-nitrobenzimidoyl chloride **9.89**, in the presence of triethylamine as base to give **9.88** (R = 4-nitrophenyl).[167] The latter reaction gave a second product, possibly due to insertion of the addend into a 6,5-bond.

9.3.4 *Formation of Pyrazolo[60]fullerenes*

Two examples of these derivatives are presently known. One results from the 1,3-dipolar reaction of a pyrazolidinium ylide **9.90** (R^1,R^2 = H or Me, R^3 = NO_2), prepared from pyrazolidinone and aldehydes, with [60]fullerene. The reaction is unsuccessful for compounds with electron-supplying R^3 due to delocalisation of the carbocation.[168] The other utilises 1,3-diphenylnitrilimine

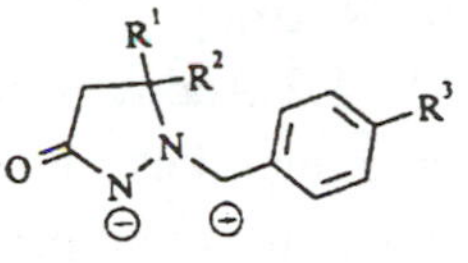

9.90

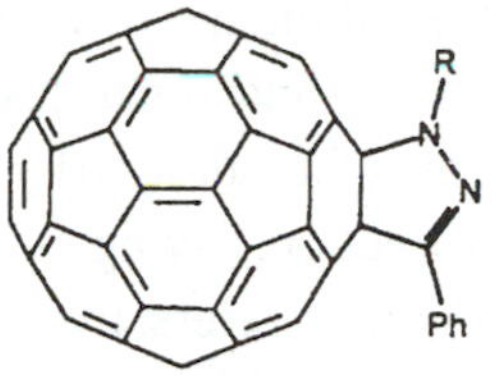

9.91

Ph-N$^-$-N = C$^+$-Ph, obtained from the reaction of triethylamine with either benzhydrazidoyl chloride or *N*-(α-chlorobenzylidene)-*N* '-phenylhydrazine to give **9.91**.[169]

9.3.5 *Formation of Isoxazolofullerenes*

1,3-Dipolar addition to [60]fullerene of nitrile oxides, R-C ≡ N$^+$-O$^-$ (obtained from RNO$_2$, PhNCO and triethylamine), produces 4,5-dihydroisoxazolo [4,5:1,2][60]fullerenes **9.92**. This reaction has been carried out successfully for R = Me, Et, Ar, CO$_2$R[170] 9-anthracenyl (and derivatives), *p*-tosyl, mesityl, CO$_2$R,[171] COPh, (CH$_2$OCH$_2$)$_2$CH$_2$OMe, and a variety of (CH$_2$)$_n$CO$_2$R groups.[172] The unsubstituted compound **9.92** (R = H) can be made by reaction of [60]fullerene with fulminic acid HON = C(Cl)-C(OH) = O (which eliminates CO$_2$ and HCl), or with nitromethane, trimethylsilyl chloride and triethylamine, followed by acid-catalysed elimination of SiMe$_3$OH from the intermediate.[173]

Reaction of **9.92** (R = H) with triethylamine in toluene at 70°C, results in ring opening to give 1-cyano-2-hydroxy-1,2-dihydro[60]fullerene **9.93**, a rare example of a monohydroxy fullerenol (see also Secs. 8.6.1 and 8.6.2).[174]

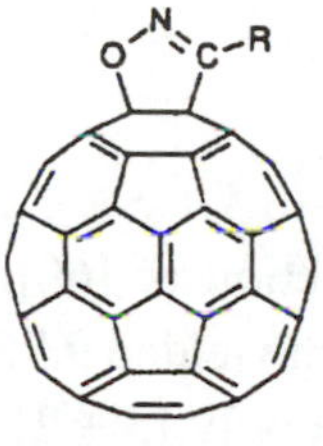

9.92

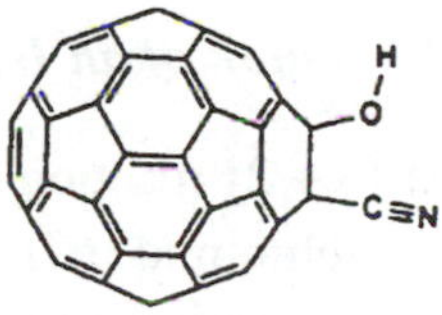

9.93

Reaction of nitrile oxides with [70]fullerene occurs more slowly than with [60]fullerenes, and regardless of the nature of the R group, 1,2-addition is favoured over 5,6-addition by a ratio of *ca.* 1.5–1.9.[175] 1,2-Addition take places in either direction, giving two isomeric products, and regardless of the nature of R, this group prefers, by a factor of *ca.* 1.5, to lie over the polar pentagon. A possible reason is that the greater curvature of the cage at the pole provides more space for the substituent R. The unsymmetrical nature of the addend also means that 5,6-addition produces an enantiomeric pair.

Isoxazolofullerenes revert back to the parent fullerene on heating with either $Mo(CO)_3$ or DIBAL-H, leading to the proposed use of the addend either for blocking, solublising, or as a tether anchor.[172]

9.3.6 *Formation of Oxazolo[60]fullerenes*

4,5-Dihydrooxazolo[4,5:1,2][60]fullerenes (**9.48**) are readily obtained on heating the *N*-carboaroxyaziridino[60]fullerene (**9.45**, R = CO_2Ar),[78] in which the aryl ring substituents are H, 4-OMe, 4-CN, 4-Br,[176] and 2,4,6-tri-t-butyl.[78]

If the corresponding ethoxy derivative (**9.45**, R = CO_2Et) is treated with either phenol/trimethylsilyl chloride[177] or BBr_3,[178] then the oxazolidin-2-ono[60]fullerene **9.94** (R = Et), is formed. (There are, however, serious discrepancies between the spectroscopic data in these two sets of work.)

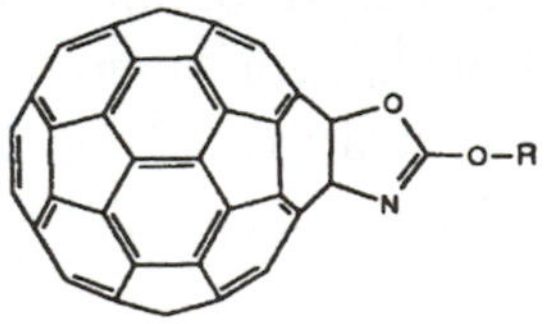

9.94

9.3.7 *Formation of an Isothiazolo[60]fullerene*

The first [3 + 2] reaction to be observed involved reaction of [60]fullerene with the sulphinimide $(CF_3)_2C = S = N$-adamantyl which yielded **9.95** accompanied by some di- and hexa-adduct;[179] this is also the only [3 + 2] reaction that produces a hexa-adduct, which is surprising in view of the addend bulk.

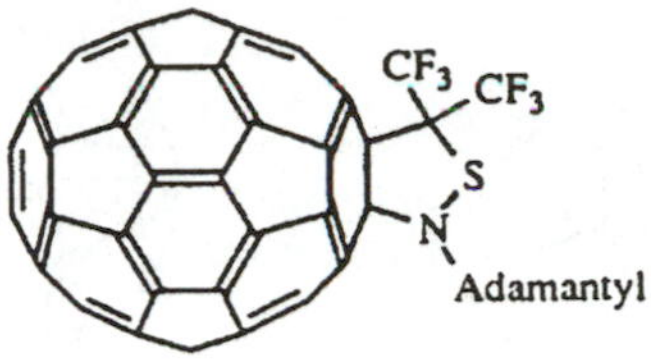

9.95

9.3.8 *Formation of a Thiazolo[60]fullerene*

Heating [60]fullerene with thioacetamide in PhCl produces 2-methyl-4,5-dihydrothiazolo[4,5:1,2][60]fullerene **9.96**,[173] and the simplicity of this reaction coupled with the ready availability of thioamides suggests that many thiazolofullerenes could become available.

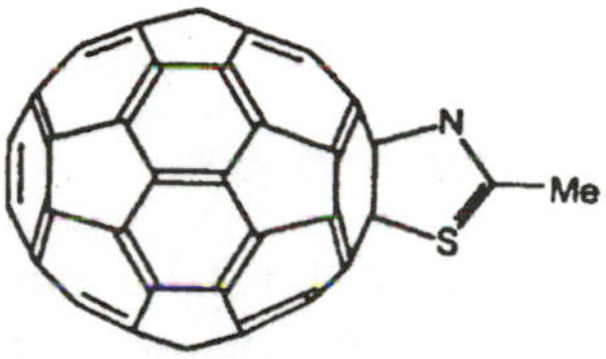

9.96

9.3.9 *Formation of Furano[60]fullerenes*

4,5-Dihydrofurano-[4,5:1,2][60]fullerenes **9.97** (R and R' variously alkyl or alkoxy) are obtained by room temperature reaction of the corresponding diketones R(CO)CH$_2$(CO)R' with [60]fullerene in the presence of piperidine.[180]

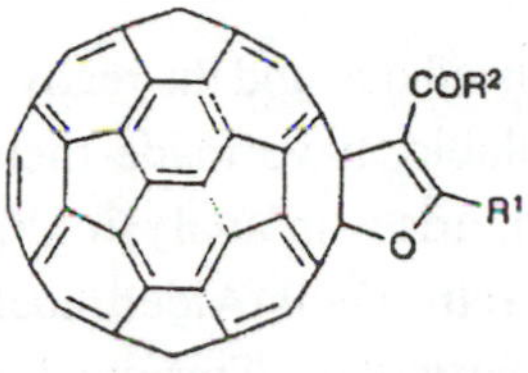

9.97

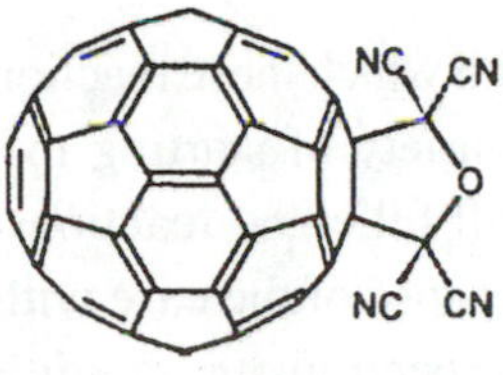

9.98

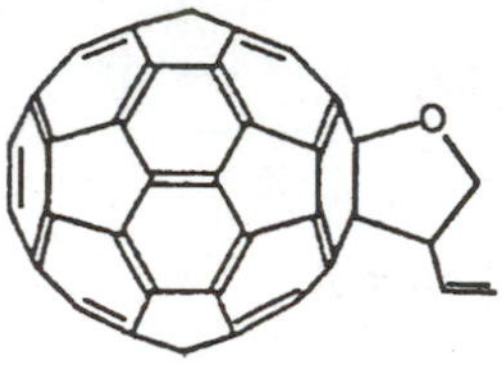

9.99 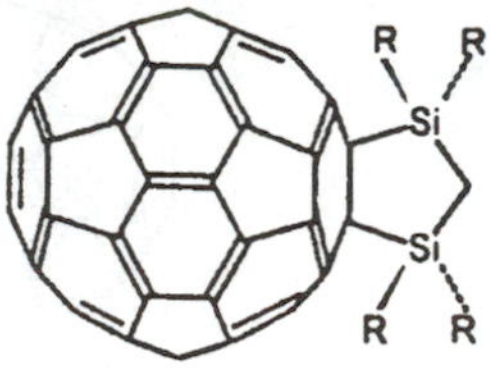   **9.100**

The formation of 2,2,5,5-tetracyanotetrahydrofurano[4,5:1,2][60]fullerene (**9.98**) by heating tetracyanoethylene oxide with [60]fullerene, proceeds probably via intermediate generation of the carbonyl ylide following opening of the 3-membered epoxide ring.[181]

In contrast to the above, 2,3-fusion results from the palladium-catalysed reaction of [60][fullerene with *cis*-HOCH$_2$CH = CHCH$_2$OCO$_2$Et giving 4-vinyl-2,3,4,5-tetrahydrofurano-[2,3:1,2][60]fullerene **9.99**.[182]

9.3.10 *Formation of Disilolanofullerenes*

Irradiation of a toluene solution of 1,1,2,2,tetramesityl-1,2-disilirane with [60]fullerene gives the disilolano[60]fullerene **9.100**, in high yield.[183] The corresponding reaction with [70]fullerene was assumed to have involved addition across the 20,21-bond giving a twisted structure (necessary to account for the NMR patterns).[184] However, this must now be considered unlikely because there no other addition across this bond is known, whereas by contrast, addition across the 7,23 positions (as is found in chlorination) accounts for all of the observed NMR patterns.

9.4 [4 + 2] Cycloadditions: Diels-Alder Reactions

The ease with which these reactions can be carried out, and the ready availability of a wide variety of starting materials available, have made them the most studied of all fullerene reactions. A problem arises in analysing the reaction products because of the ease with which the retro Diels-Alder reaction occurs under mass spectrometry conditions (EI in particular). This can be overcome by addition to the addend thereby preventing the reverse reaction from occurring,[185] and also by greatly enhancing the stability of the products through the introduction of aromaticity. This is described in the next section.

9.4.1 *Reactions of o-Quinodimethane and Derivatives*

This technique (which involves a specific reaction of 1,3-butadienes, described more generally in Sec. 9.4.3) was introduced by Müllen and coworkers[186] and has been greatly utilised. The basic method (Fig. 9.13) involves reaction with 1,2-bis(bromomethyl)benzene with KI (to produce 1,4-elimination of HBr) in THF in the presence of a crown ether phase-transfer catalyst, (necessary because of the low solubility of [60]fullerene in THF). Some diaddition occurs, and a dumbbell structure can also be obtained (see Sec. 12.6) by use of a 1,2,4,5-tetra(bromomethyl)benzene. Because the reaction products gain aromaticity, there is a substantially reduced tendency to undergo the retro Diels-Alder reaction, so that the products are stable.

Numerous examples of this reaction include formation of derivatives having carboxyl, ester, amide groups,[187,188] alkoxy groups, crown ethers,[189] etc. as substituents in the benzene ring, with naphthalene, pyrene, phenanthrene[190] in place of the benzene ring, and with hydroxyquinone derivatives.[191] The reaction has been used to prepare 1,2-, 5,6- and 7,21-derivatives of [70]fullerene in 24, 10 and 1–2% respective yields, and also to provide the first investigation of the regiochemistry of [76]fullerene.[192] This led to the formation of at least six isomers, the major one being characterised as probably the 2,3-isomer, the next most abundant being the 1,6-isomer. This was interpreted in terms of maximum local curvature in the molecule,[192] and also agrees quite well with calculated densities of the respective π-bonds.[59]

Other examples of the reaction include having the 1,2-bis(bromomethyl) benzene groups linked by means of -O(CH$_2$)nO- tethers leading (after cleavage

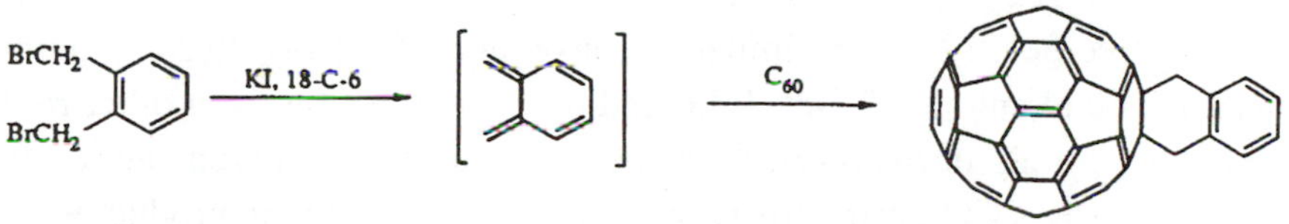

Fig. 9.13 Formation of stable Diels-Alder adducts through the use of *o*-quinodimethane.

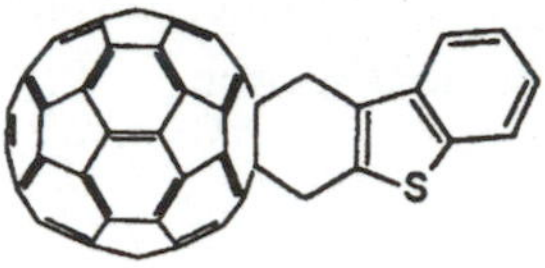

9.101 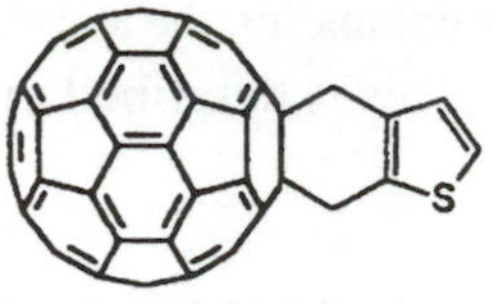   9.102 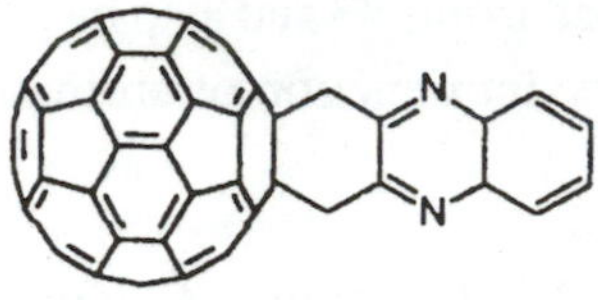   9.103

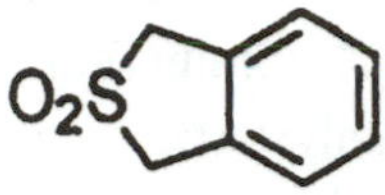

9.104 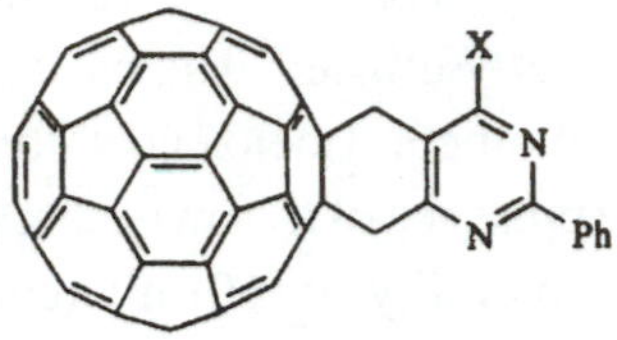   9.105 (X = OMe, piperazinyl).

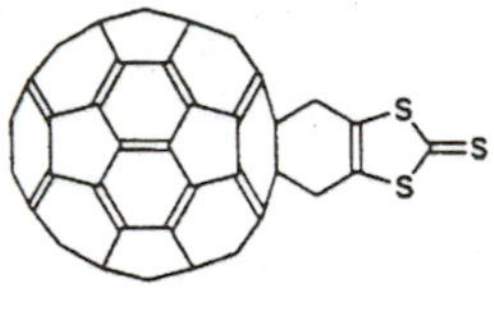

9.106 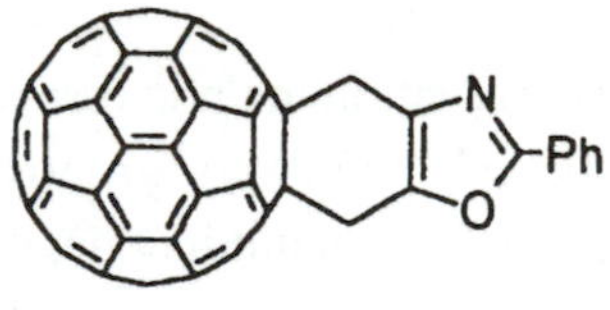   9.107 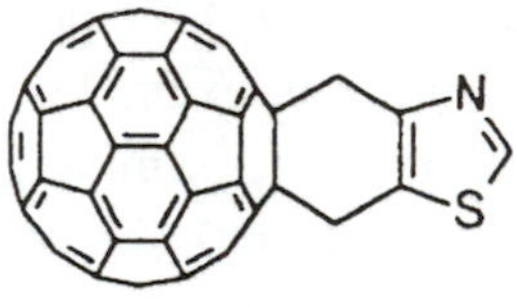   9.108

of the tether) to the formation of either *cis*-2-, *cis*-3- and *e*-bis-phenolic derivatives.[193] Variations of the reaction have been used to prepare further dumbbell structures,[194] and heterocyclic derivatives such as **9.101**,[195] **9.102** and **9.103**.[196]

The *o*-quinodimethane intermediates can also be prepared by converting the bis-bromomethyl starting material into a sulphone, e.g. **9.104**, followed by thermal extrusion of SO_2. Various derivatives in which electron donor groups replace the amino group have been synthesised in this way,[197] and also heterocyclic compounds such as **9.105–9.108**,[196,198,199] and polymers (see Sec. 12.6).[200] Similarly, conversion of the bis-dibromomethyl compounds into sultines **9.109**, followed by SO_2 extrusion, is another route to making various derivatives.[201] Another variation employed the ketone **9.110** which loses CO during the reaction to give the aromatic product **9.111**, and so gain stability.[202] A unique reaction involves photoconversion of

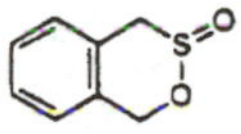

9.109

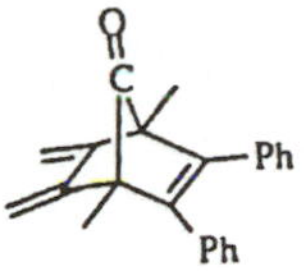

9.110

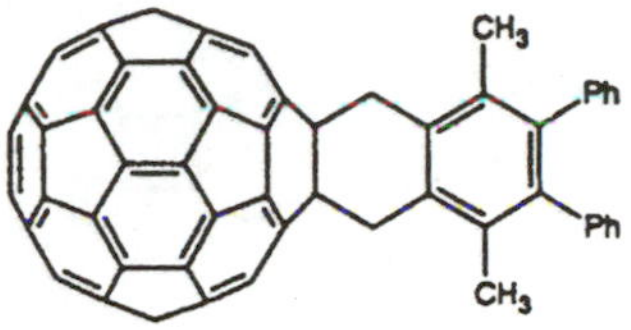

9.111

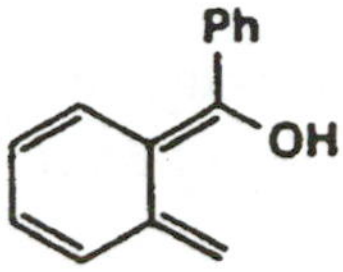

9.112

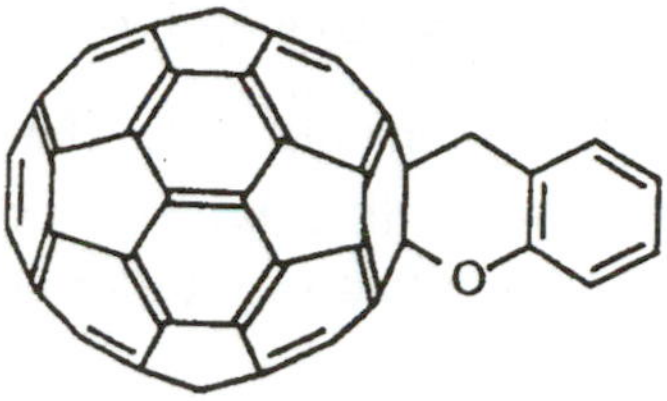

9.113

2-methylbenzophenone (1,5-hydrogen shift) into the *o*-quinodimethane derivative **9.112**, which then adds to [60]fullerene. The product is unstable and undergoes ring opening to form 1-(2-benzoylphenyl)methyl-1,2-dihydro[60]fullerene.[203]

Other methods have been used to introduce a heteroatom in place of one of the methylene groups, so leading to products such as **9.113**.[204]

Yet another variation involves the use of isobenzofuran and derivatives **9.114**, which are relatively unstable 10π aromatics and become converted into the more stable 6π benzenoid derivatives **9.115** following addition.[114,205] A rather less stable starting material is isoindene **9.116**, a 10π system through the involvement of C-H hyperconjugation, the addition of which produces **9.117**.[206]

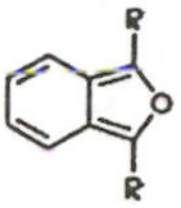

9.114

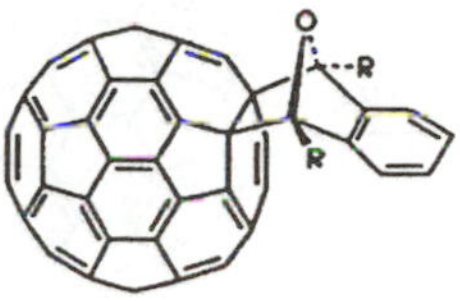

9.115

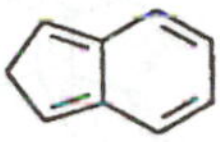

9.116

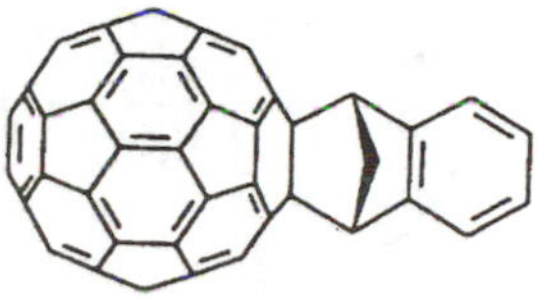

9.117

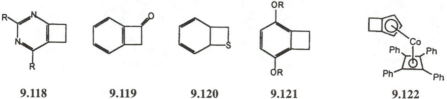

| 9.118 | 9.119 | 9.120 | 9.121 | 9.122 |

9.4.2 *Reactions of Cyclobutane Derivatives*

Thermal ring-opening of arenocyclobutanes produces *o*-quinodimethanes which can then add to [60]fullerene as above.[207] Some examples of precursors that have been used in this way are shown in **9.118–9.122**.[208]

9.4.3 *Addition of Buta-1,3-Diene Derivatives*

The most basic reaction of this type involves substituted buta-1,3-dienes, e.g. Fig. 9.14. The reaction can give products with six addends located at the octahedral sites of the fullerene, and the intermediate compounds can be isolated.[209] It is noteworthy that calculations on this reaction indicate that differential solvation may be responsible for the variable reactivity difference between [60]- and [70]-fullerenes towards addition in general; [70]fullerene is predicted to be the more reactive in the gas phase and the less reactive in solution.[210] Reactions carried out both with and without solvent support this hypothesis.[211]

The strength of the electron withdrawal by the fullerene cage is such, that addition occurs even with dienes having electron-withdrawing groups (CO_2Et, $COMe$, CN, SO_2Ph, NO_2) attached. Moreover, cycloreversion of the products is apparently inhibited by these substituents (but see also Sec. 9.4.5).[212] The

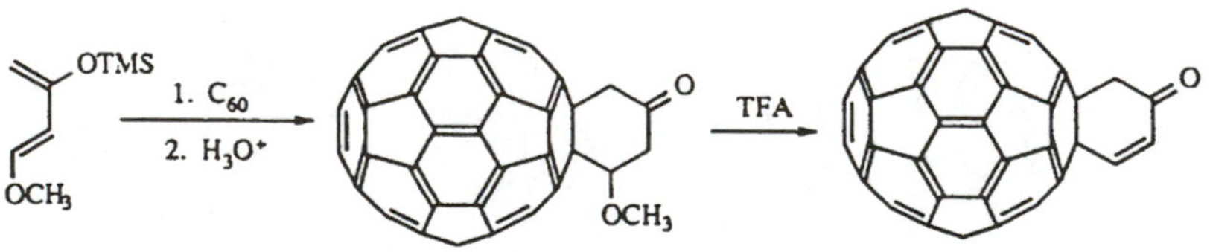

Fig. 9.14 Reaction of [60]fullerene with buta-1,3-diene and derivatives.

cycloreversion is a major problem in EI mass spectroscopic analysis of cycloaddition products, and one solution may be (in principle at least) to use electrospray. This was demonstrated by using a crown ether (particularly suitable for electrospray) linked via a phenyl ring to butadiene which was then cycloadded to [60]fullerene.[213] This approach may however be too complicated for general application.

Reaction with both 1- and 2-trimethylsilyloxybuta-1,3-dienes can be employed for the formation of ketonic products and further derivatives such as alcohols, useful precursors for formation of biologically important molecules.[212,214,215] For example, addition with the 2-isomer leads to the ketone **9.123** which can be reduced to the corresponding water-soluble alcohol and this in turn esterified.[214] The 1-isomer has been used to prepare the alcohol **9.124** which may be converted *via* the intermediate diene **9.125**, to the bis-homofullerene **9.126** through consecutive [4 + 4] and retro [2 + 2 + 2] reactions.[215] Notably, formation of an η^2 derivative of **9.126** with cyclopentadienylcobalt(I) was accompanied by bond-cleavage producing a nine-membered hole on the cage surface.[215]

Hydroxy groups in the addend can also be introduced by photochemical oxidation of 1,2-cyclohex-3'-eno[60]fullerene (produced by cycloaddition of buta-1,3-diene) which yields 1,2-(5'-hydroxycyclohex-3'-eno)[60]fullerene.[214] A carbonate formed by subsequent esterification is very soluble, and the addend can be readily removed after further additions carried out elsewhere on the cage.

Buta-1,3-diene can be cycloadded to opposite sides of [60]fullerene through being attached to the ends of a tether (see also Sec. 9.1.1.5). The addends are thus equatorial relative to the tether anchor which leave three mutually equatorial positions for further addtion; in this way tetra- and hexa-adducts have been prepared.[216]

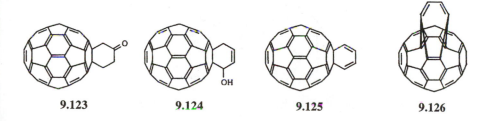

9.123   **9.124**   **9.125**   **9.126**

A cycloadduct of [60]fullerene with buta-1,3-diene linked to a nickel phthalocyanine moiety has been prepared in anticipation that the electron-acceptor properties of the fullerene would influence the electronic properties of the phthalocyanine, but this proved not to be the case.[217] The problem is a fundamental (and generally unrecognised) one, namely that linkage of any addend to the cage is necessarily through sp^3-hybridised carbons and this effectively prevents any conjugative interactions, regardless of any conjugation in the addend. So far, only indirect (periconjugation) between an addend and the cage has been observed (see Sec. 9.1.1).

High pressure is necessary to facilitate some cycloadditions. For example, although [60]fullerene undergoes addition with both 2,3-dimethylene-1,4-dioxane, high pressure is needed to bring about reaction with 4,5-dimethylene-2,2-dimethyldioxolane, which is less reactive due to a less favourable geometry. These compounds provide a route to formation of ketonic e.g. acyloin derivatives.[218] High pressure brings about reaction between [60]fullerene and several disubstituted tropones, e.g. **9.127**, addition to the addend occurring where indicated;[219] reaction with [70]fullerene appears to be slower. It also aids reaction between [60]fullerene and cycloheptatriene (two products are obtained due to isomerisation of the triene to norcaradiene),[220] and reaction between [60]fullerene and 2-cycloalkenones and their acetals to give **9.128**;[221] this reaction can also be used to prepare chiral [60]fullerene norbornan-2-one derivatives through the use of a chiral auxilliary in the starting acetal.[222]

Heterocyclic derivatives of fullerenes may also be made by replacing a heteroatom in the butadiene chain, e.g. using the 2-azabuta-1,3-diene derivative **9.129**, and the *N*-acylthioacrylamide **9.130**.[223]

9.127 (R and R′ variously, H, Br, Cl, alkyl, OMe, OAc, CO_2 alkyl)

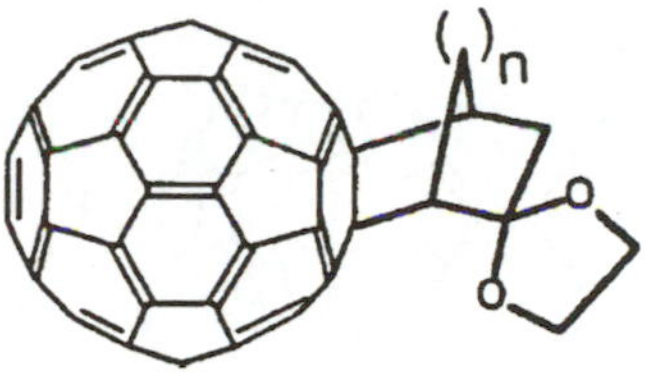

9.128 (*n* = 1,2).

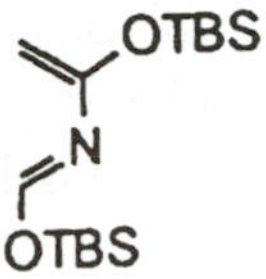

9.129

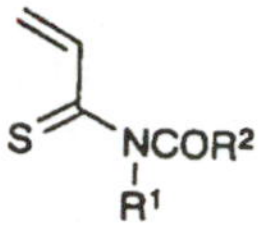

9.130

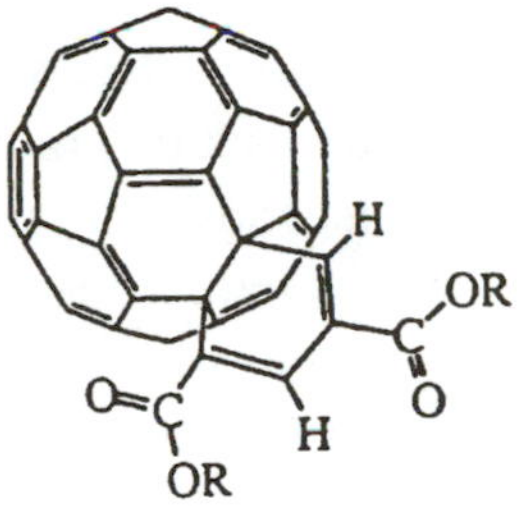

9.131

Lastly, compounds **9.131** can be prepared by reaction of [60]fullerene with two molecules of alkyl propynoates in the presence of tricyclohexylphosphine in a reaction described as formally involving a [2 + 2 + 2] mechanism.[224] However, the mechanism involves initial combination of the two alkynoates to give a dipolar butadiene which then adds to the fullerene and so may strictly be regarded as a [4 + 2] cycloaddition.

9.4.4 *Addition of Anthracene*

Addition (across the 9,10-positions) of anthracene to [60]fullerene takes place very readily, is thermally reversible, and has been well documented.[41,114,225] Diaddition also occurs readily giving **9.133** and is unusual in taking place at directly opposite sides of the cage,[41] presumably due to steric hindrance towards addition at other sites. Thermal disproportionation in the solid monoadduct also occurs to give **9.133** and [60]fullerene, apparently favoured here by the crystal structure, believed to be consist of stacks of anthraceno[60]fullerenes in which the cage face of one molecule is closely accommodated by the concave face of the anthracene addend on the next.[226]

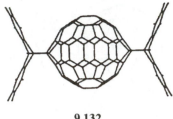

9.132

The bulk of the anthracene addend and its ready removal, means that this is a useful blocking group that can be used to restrict the number and location of other addends, and also to activate particular sites towards other additions. Dimethylanthracene (which can readily be removed by $O_2/h\nu$, or by dimethyl acetylenedicarboxylate) has been used in this way, and leads for example to a 120-fold increase in yield of the product of octahedral addition of diethyl malonate addends.[227]

Addition to [70]fullerene itself has not been reported, though addition to $C_{70}Ph_8$ occurs across the most reactive double bond (C48–C49) in the molecule; this is the only double bond present in a pentagonal ring, and is also subject to spontaneous oxidation (see Sec. 8.6.2).[228]

9.4.5 *Addition of Cyclopentadiene, Furan and Derivatives*

The ready availability of cyclopentadiene resulted in some of the earliest cycloadditions to fullerenes.[41,114,229] With [60]fullerene, up to six addends can attach to the cage, most probably at the octahedral sites.[185] The problem of EI mass spectrometric analysis, which under EI condition results in cycloreversion, can be overcome by either hydrogenation or bromination of the addend. Use of dilute solutions of bromine ensured selective bromination of the addend rather than the cage (which requires concentrated bromine solutions). Likewise, selective hydrogenation of the addend is achieved by using hydrogen and Adam's catalyst, a condition which will not hydrogenate the cage.

Reaction of both [60]- and [70]fullerenes with pentamethylcyclopentadiene gives products that are less prone to cycloreversion than those obtained with the unmethylated species.[230] This is consistent with the expected greater ease of formation arising from increased electron supply in the diene. Addition to

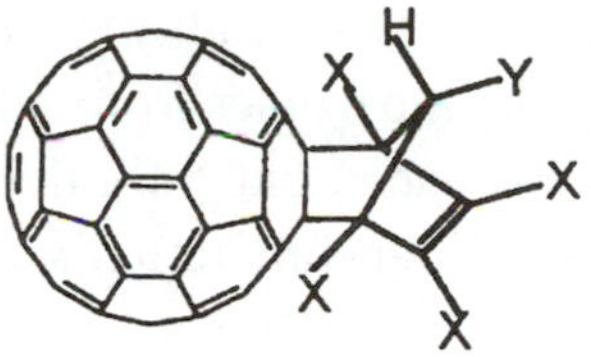

9.133

[70]fullerene (which takes place almost exclusively across the 1,2-bond) occurs less readily than addition to [60]fullerene. This is attributable to the lower electrophilicity of [70]fullerene, confirmed by the relative positions of the resonance for the single hydrogen in the proton NMR of the respective adducts (more downfield in the [60]fullerene adduct). This hydrogen also points towards the cage in both adducts **9.133** (X, Y = Me), which provided an early indication of the importance of steric effects in fullerene chemistry.

The importance of steric effects is also shown in the bromination and epoxidation of the double bond in the addend resulting from reaction of [60]fullerene with cyclopentadiene. Both the bromines and the oxygen point away from the cage, and most notably, addition of bromine to the double bond occurs *cis* rather than *trans*.[231] Steric effects are also apparent in the reaction of $(\eta^5\text{-}C_5H_5ZrCl)_nC_{60}H_n$ with N-bromosuccinimide which gives **9.133** (X = H, Y = Br). Likewise the reaction with peracid gives the corresponding **9.133** (X = H, Y = OH).[232]

Further chemistry of the cyclopentadiene adducts is largely unexplored. The epoxide described above eliminates CO under EI mass spectrometry conditions indicating a possible route to a 1,3-cyclobuta[60]fullerene,[231] and reaction of **9.133** (X, Y = H) with 3,6-di(2-pyridyl)-*s*-tetrazine gives products such as **9.134**.[233]

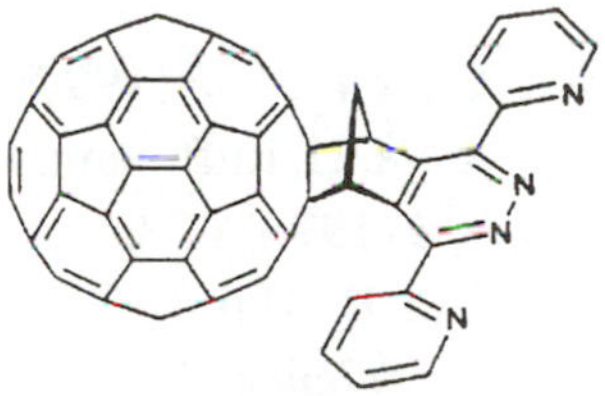

9.134

Addition of furan should occur less readily that of cyclopentadiene because it has greater aromaticity, which becomes lost in the product. No measure of the relative reactivities is available, but [60]fullerene gives an adduct with furfural,[234] and also with a furan-functionalised polymer resin, giving a polymeric product.[235]

9.5 [6 + 2] Reactions

Only one reaction of this type is known at present. The photochemical reaction between [60]fullerene and *N*-ethoxycarbonylazepine gives two products **9.135** and **9.136** (E = CO_2Et), in *ca.* 4:1 ratio, arising from [4 + 2] and [6 + 2] reactions.[236]

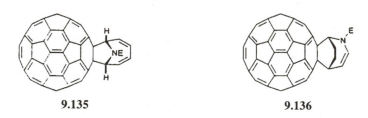

9.135   **9.136**

9.6 [8 + 2] Reactions

The only example (see also Sec. 9.3.1)involves the reaction of 8-methoxyheptafulvene with [60]fullerene to give the tetrahydroazulenofullerene **9.83**.[147] This process provides one route to preparing chiral fullerene derivatives, and further additions on the cycloheptatriene unit are possible.

References

1. S. Sliwa, *Fullerene Sci. & Technol.*, **5** (1997) 1133.
2. J. Averdung, G. Torres-Garcia, H. Luftmann, I. Schlachter and J. Mattay, *Fullerene Sci. & Technol.*, **4** (1996) 633.
3. F. Diederich, L. Isaacs and D. Philp, *Chem. Soc. Rev.*, (1994) 243; F. Diederich and C. Thilgen, *Science*, **271** (1996) 317; M. Prato, *J. Mater. Chem.*, **7** (1997) 1097.

4. J. M. Wood, B. Kahr, S. H. Hoke, L. Dejarme, R. G. Cooks and D. Ben-Amotz, *J. Am. Chem. Soc.*, **113** (1991) 5907.
5. P. R. Birkett, A. D. Darwish, H. W. Kroto, G. J. Langley, R. Taylor and D. M. Walton, *J. Chem. Soc., Perkin Trans.* 2, (1995) 511.
6. A. D. Darwish, A. K. Abdul-Sada, G. J. Langley, H. W. Kroto, R. Taylor and D. M. Walton, *J. Chem. Soc., Perkin Trans.* 2, (1995) 2359.
7. A. G. Avent *et al.*, *J. Chem. Soc., Perkin Trans.* 2, (1998) 1319.
8. T. Suzuki, Q. Li, K. C. Khemani, F. Wudl and Ö. Almarsson, *Science*, **254** (1991) 1186.
9. T. Suzuki, Q. Li, K. C. Khemani and F. Wudl, *J. Am. Chem. Soc.*, **114** (1992) 7301.
10. F. Wudl, *Acc. Chem. Res.*, **25** (1992) 157.
11. A. B. Smith *et al.*, *J. Am. Chem. Soc.*, **115** (1993) 5829.
12. J. Osterodt, M. Nieger and F. Vögtle, *J. Chem. Soc., Chem. Commun.*, (1994) 1607; H. L. Anderson, C. Boudon, F. Diederich, J. Gisselbrecht, M. Gross and P. Seiler, *Angew. Chem. Intl. Edn. Engl.*, **33** (1994) 1628.
13. A. B. Smith *et al.*, *J. Am. Chem. Soc.*, **117** (1995) 5492.
14. R. A. J. Janssen, J. C. Hummelen and F. Wudl, *J. Am. Chem. Soc.*, **117** (1995) 544.
15. M. Eiermann, F. Wudl, M. Prato and M. Maggini, *J. Am. Chem. Soc.*, **116** (1994) 8364.
16. R. González, J. C. Hummelen and F. Wudl, *J. Org. Chem.*, **60** (1995) 2618; J. C. Hummelen, B. W. Knight, F. LePeq, F. Wudl, J. Yao and C. L. Wilkins, *J. Org. Chem.*, **60** (1995) 532.
17. M. Prato *et al.*, *J. Am. Chem. Soc.*, **115** (1993) 8479.
18. H. Tokuyama, M. Nakamura and E. Nakamura, *Tetrahedron Lett.*, **34** (1993) 7429; H. Tokuyama, H. Isobe and E. Nakamura, *Bull. Chem. Soc. Jpn.*, **68** (1995) 935.
19. Z. Li and P. B. Shevlin, *J. Am. Chem. Soc.*, **119** (1997) 1149.
20. E. Vogel, *Pure Appl. Chem.*, **54** (1982) 1015.
21. L. Isaacs, A. Wehrsig and F. Diederich, *Helv. Chim. Acta*, **76** (1993) 1231.
22. A. Skiebe and A. Hirsch, *J. Chem. Soc., Chem. Commun.*, (1994) 335.
23. F. Diederich, C. Dietrich-Buchecker, J. Nierengarten and J. Sauvage, *J. Chem. Soc., Chem. Commun.*, (1994) 781.
24. I. Lamparth, G. Schick and A. Hirsch, *Liebigs Ann. Chem.*, (1997) 253.

25. L. Isaacs and F. Diederich, *Helv. Chim. Acta*, **76** (1993) 2454.

26. M. Keshavarz-K, B. Knight, R. C. Haddon and F. Wudl, *Tetrahedron*, **52** (1996) 5149.

27. J. Osterodt, M. Nieger, P. Windscheif and F. Vögtle, *Chem. Ber.*, **126** (1993) 2331; J. Osterodt and F. Vögtle, *Fullerene Sci. & Technol.*, **4** (1996) 729.

28. F. Arias, L. A. Godinez, S. R. Wilson, A. E. Kaifer and L. Echegoyen, *J. Am. Chem. Soc.*, **118** (1996) 6086.

29. U. Jonas *et al.*, *Chem. Eur.*, **1** (1995) 243.

30. S. H. Friedman, D. L. DeCamp, R. P. Sijbesma, G. Srdanov, F. Wudl and G. L. Kenyon, *J. Am. Chem. Soc.*, **115** (1993) 6510.

31. R. F. Schinazi, R. P. Sijbesma, G. Srdanov, C. L. Hill and F. Wudl, *Antimicrob. Agents Chemother.*, **37** (1993) 1707.

32. M. Prato, A. Bianco, M. Maggini, G. Scorrano, C. Toniolo and F. Wudl, *J. Org. Chem.*, **58** (1993) 5578.

33. C. Zhu, Y. Xu, Y. Liu and D. Zhu, *J. Org. Chem.*, **62** (1997) 1996.

34. A. Vasella, P. Uhlmann, C. A. A. Waldruff, F. Diederich and C. Thilgen, *Angew. Chem. Intl. Edn. Engl.*, **31** (1992) 1388.

35. Y. Z. Anz, Y. Rubin, C. Schaller and S. W. McElvany, *J. Org. Chem.*, **59** (1994) 2927; H. L. Anderson, R. Faust, Y. Rubin and F. Diederich, *Angew. Chem. Intl. Edn. Engl.*, **33** (1994) 1366; R. González, J. C. Hummelen and F. Wudl, *J. Org. Chem.*, **60** (1995) 2618; Z. Li, K. H. Bouhadir and P. B. Shevlin, *Tetrahedron Lett.*, **37** (1996) 4651.

36. T. Ishida, K. Shinozuka, M. Kubata, M. Ohashi and T. Nogami, *J. Chem. Soc., Chem. Commun.*, (1995) 1841.

37. M. Kawasumi, T. Ishida and T. Nogami, *Fullerene Sci. & Technol.*, **4** (1993) 357.

38. K. Komatsu, Y. Murata, A. Miyabo, K. Takeuchi and T. S. M. Wan, *Fullerene Sci. & Technol.*, **1** (1993) 231.

39. M. T. H. Liu, Y. N. Romashin, M. Kubata and M. Ohashi, *Org. Mass Spectrom.*, **29** (1994) 391.

40. W. W. Win *et al.*, *J. Org. Chem.*, **59** (1994) 5871.

41. M. Tsuda, T. Ishida, T. Nogami, S. Kurono and M. Ohashi, *Tetrahedron Lett.*, **34** (1993) 6911; *Fullerene Sci. & Technol.*, **1** (1993) 275; **2** (1994) 155; *J. Chem. Soc., Chem. Commun.*, (1993) 1296.

42. R. González, F. Wudl, D. L. Pole, P. K. Sharma and J. Warkentin, *J. Org. Chem.*, **61** (1996) 5837.

43. C. Eaborn, *J. Chem. Soc.*, (1956) 4858.

44. T. Ishida, T. Furudate, T. Nogami, M. Kubota, T. Hirano and M. Ohashi, *Fullerene Sci. & Technol.*, **3** (1995) 399; A. M. Benito, A. D. Darwish, H. W. Kroto, M. F. Meidine, R. Taylor and D. R. M. Walton, *Tetrahedron Lett.*, **37** (1996) 1085.

45. P. L. Boulas, Y. Zuo and L. Echegoyen, *Chem. Commun.*, (1996) 1547.

46. J. Osterodt and F. Vögtle, *Chem. Commun.*, (1996) 547.

47. H. Yamaguchi, S. Murata, T. Akasaka and T. Suzuki, *Tetrahedron Lett.*, **38** (1997) 3529.

48. F. Wudl, T. Suzuki and M. Prato, *Synthetic Metals*, **59** (1993) 297; M. Eiermann *et al.*, *Angew. Chem. Intl. Edn. Engl.*, **34** (1995), 1591.

49. T. Ohno, N. Martin, B. Knight, F. Wudl, T. Suzuki and H. Yu, *J. Org. Chem.*, **61** (1996) 1306.

50. T. Benicori, E. Brenna, F. Sannicolo, L. Trimarco, G. Zotti and P. Sozzani, *Angew. Chem. Intl. Edn. Engl.*, **35** (1996) 648.

51. J. E. Chateauneuf, *J. Am. Chem. Soc.*, **117** (1995) 2677.

52. J. Xu, Y. Li, D. Zheng, J. Yang, Z. Mao and D. Zhu, *Tetrahedron Lett.*, **44** (1997) 6613.

53. H. J. Bestman, D. Hadawi, T. Röder and C. Moll, *Tetrahedron Lett.*, **35** (1994) 9017; Y. Wang, J. Cao, D. I. Schuster and S. R. Wilson, *Tetrahedron Lett.*, **36** (1995) 6843.

54. Y. Wang, D. I. Schuster, S. R. Wilson and C. J. Welch, *J. Org. Chem.*, **61** (1996) 5198.

55. C. Bingel, *Chem. Ber.*, **126** (1993) 1957; C. Bingel and H. Schiffer, *Liebigs Ann. Chem.*, (1995) 1551.

56. X. Camps and A. Hirsch, *Chem. Commun.*, (1997) 1595.

57. J. Nierengarten and J. Nicoud, *Tetrahedron Lett.*, **44** (1997) 7737.

58. A. Hermann and F. Diederich, *Helv. Chim. Acta*, **79** (1996) 1741; *J. Chem. Soc.*, *Perkin Trans. 2*, (1997) 1679.

59. R. Taylor, *J. Chem. Soc.*, *Perkin Trans. 2*, (1993) 813.

60. C. A. Merlic and H. D. Bendorf, *Tetrahedron Lett.*, **35** (1994) 9529.

61. L. Shu, G. Wang, S. Wu and H. Wu, *J. Chem. Soc.*, *Chem. Commun.*, (1995) 367.

62. A. Herrmann, M. Rüttiman, C. Thilgen and F. Diederich, *Helv. Chim. Acta*, **78** (1995) 1673.
63. R. Taylor and D. M. Walton, *Nature*, **363** (1993) 685.
64. A. Hirsch, I. Lamparth and H. R. Karfunkel, *Angew. Chem. Intl. Edn. Engl.*, **33** (1994) 437.
65. A. Hirsch, I. Lamparth, T. Grösser and H. R. Karfunkel, *J. Am. Chem. Soc.*, **116** (1994) 9385.
66. B. Gross, V. Schurig, I. Lamparth, A. Herzog, F. Djojo and A. Hirsch, *Chem. Commun.*, (1997) 1117.
67. I. Lamparth and A. Hirsch, *J. Chem. Soc.*, *Chem. Commun.*, (1994) 1727.
68. S. R. Wilson and Q. Li, *Tetrahedron Lett.*, **36** (1995) 5707.
69. I. Lamparth, C. Maichle-Mössmer and A. Hirsch, *Angew. Chem. Intl. Edn. Engl.*, **34** (1995) 1607.
70. I. Lamparth, A. Herzog and A. Hirsch, *Tetrahedron*, **52** (1996) 5065.
71. J. Nierengarten, V. Gramlich, F. Carullo and F. Diederich, *Angew. Chem. Intl. Edn. Engl.*, **35** (1996) 2101.
72. F. Cardullo *et al.*, *Helv. Chim. Acta*, **80** (1997) 343.
73. R. Schwenninger, T. Müller and B. Kräutler, *J. Am. Chem. Soc.*, **119** (1997) 9317.
74. L. Isaacs, P. Seiler and F. Diederich, *Angew. Chem. Intl. Edn. Engl.*, **34** (1995) 1466; F. Carullo, L. Isaacs, F. Diederich, J. Gisselbrecht, C. Boudon and M. Gross, *Chem. Commun.*, (1996) 797; P. Timmerman, L. E. Witschel, F. Diederich, C. Boudon, J. Gisselbrecht and M. Gross, *Helv. Chim. Acta*, **79** (1996) 6; R. F. Haldimann, F. Klärner and F. Diederich, *Chem. Commun.*, (1997) 237; L. Isaacs, F. Diederich and R. F. Haldimann, *Helv. Chim. Acta*, **80** (1997) 317. J. Nierengarten *et al.*, *Helv. Chim. Acta*, **80** (1997) 293.
75. C. Boudon *et al.*, *Helv. Chim. Acta*, **78** (1995) 1334.
76. M. Prato, Q. C. Li, F. Wudl and V. Lucchini, *J. Am. Chem. Soc.*, **115** (1993) 1148.
77. T. Ishida, K. Tanaka and T. Nogami, *Chem. Lett.*, (1994) 561; C. J. Hawker, K. L. Wooley and J. M. J. Frechet, *J. Chem. Soc.*, *Chem. Commun.*, (1994) 925; C. J. Hawker, P. M. Saville and J. M.White, *J. Org. Chem.*, **59** (1994) 3503; M. Yan, S. X. Cai and J. F. W. Keana, *J. Org. Chem.*, **59** (1994) 5951; L. Shiu, K. Chien, T. Liu, T. Lin, G. Her and T. Luh, *J. Chem. Soc.*, *Chem. Commun.*, (1995) 1159; T. Grösser,

M. Prato, V. Lucchini, A. Hirsch and F. Wudl, *Angew. Chem. Intl. Edn. Engl.*, **34** (1995) 1343.

78. M. R. Banks, J. I. G. Cadogan, P. K. G. Hodgson, P. R. R. Langridge-Smith and D. W. H. Rankin, *J. Chem. Soc., Chem. Commun.*, (1994) 1365.

79. A. B. Smith and H. Tokuyama, *Tetrahedron*, **52** (1996) 5257; G. Schick, T. Grösser and A. Hirsch, *J. Chem. Soc., Chem. Commun.*, (1995) 2289.

80. J. Averdung, H. Luftmann, J. Mattay, K. Claus and W. Abraham, *Tetrahedron Lett.*, **36** (1995) 2543.

81. S. Kuwashima, M. Kubota, K. Kushida, T. Ishida, M. Ohashi and T. Nogami, *Tetrahedron Lett.*, **35** (1994) 4371.

82. M. R. Banks *et al.*, *Tetrahedron Lett.*, **35** (1994) 9067.

83. J. Zhou, A. Rieker, T. Grösser, A. Skiebe and A. Hirsch, *J. Chem. Soc., Perkin Trans. 2*, (1997) 1.

84. N. Wang, J. Li, D. Zhu and T. K. Chan, *Tetrahedron Lett.*, **36** (1995) 431.

85. M. R. Banks *et al.*, *J. Chem. Soc., Chem. Commun.*, (1995) 885.

86. J. Averdung, C. Wolff and J. Mattay, *Tetrahedron Lett.*, **37** (1996) 4683.

87. M. R. Banks *et al.*, *Chem. Commun.*, (1996) 507.

88. B. Nuber and A. Hirsch, *Fullerene Sci. & Technol.*, **4** (1996) 715; C. Bellavia-Lund and F. Wudl, *J. Am. Chem. Soc.*, **119** (1997) 943.

89. Y. N. Yamakoshi, T. Yagami, S. Sueyoshi and N. Miyata, *J. Org. Chem.*, **61** (1996) 7236.

90. N. Jagerovic, J. Elguero and J. Aubagnac, *Tetrahedron*, **52** (1996) 6733.

91. J. C. Hummelen, M. Prato and F. Wudl, *J. Am. Chem. Soc.*, **117** (1995) 7003.

92. R. Taylor, *J. Chem. Res. (S)*, (1987) 178.

93. J. C. Hummelen, M. Keshavarz-K, J. L. J. van Dongen, R. A. J. Janssen, E. W. Meijer and F. Wudl., *Chem. Commun.*, (1998) 281.

94. L. Shiu, K. Chien, T. Liu, T. Lin, G. Her and T. Luh, *J. Chem. Soc., Chem. Commun.*, (1995) 1159; G. Dong, J. Li and T. Chan, *J. Chem. Soc., Chem. Commun.*, (1995) 1725; C. K. Shen, K. Chien, C. Juo, G. Her and T. Luh, *J. Org. Chem.*, **61** (1996) 9242.

95. G. Schick, A. Hirsch, H. Mauser and T. Clark, *Chem. Eur. J.*, **2** (1996) 935.

96. R. Taylor, *Proc. Electrochem. Soc.*, **97–14** (1997) 281.

97. T. Grösser, M. Prato, V. Lucchini, A. Hirsch and F. Wudl, *Angew. Chem. Intl. Edn. Engl.*, **34** (1995) 1343.

98. I. Lamparth, B. Nuber, G. Schick, A. Skiebe, T. Grösser and A. Hirsch, *Angew. Chem. Intl. Edn. Engl.*, **34** (1995) 2257.

99. V. N. Bezmelnitsin, A. V. Eletskii, N. G. Schepetov, A. G. Avent and R. Taylor, *J. Chem. Soc., Perkin Trans. 2*, (1997) 683.

100. F. Diederich *et al.*, *Science*, **252** (1991) 548.

101. R. Taylor and N. J. Tower, unpublished work.

102. K. M. Creegan *et al.*, *J. Am. Chem. Soc.*, **114** (1992) 1103.

103. S. W. McElvany, J. H. Callahan, M. M. Ross, L. D. Lamb and D. R. Huffman, *Science*, **260** (1993) 1632.

104. A. B. Smith *et al.*, *J. Org. Chem.*, **61** (1996) 1904.

105. L. Juha, V. Hamplová, J. Kodymoná and O. Spalek, *J. Chem. Soc., Chem. Commun.*, (1994) 2437.

106. Y. Elemes *et al.*, *Angew. Chem. Intl. Edn. Engl.*, **31** (1992) 351.

107. T. Hamano, T. Mashino and M. Hiroba, *J. Chem. Soc., Chem. Commun.*, (1995) 1537.

108. R. W. Murray and K. Iyanar, *Tetrahedron Lett.*, **38** (1997) 335.

109. W. A. Kalsbeck and H. H. Thorp, *J. Electroanal. Chem.*, **314** (1991) 363.

110. D. Heymann and L. P. F. Chibante, *Rec. Trav. Chim.*, **112** (1993) 531, 639; *Chem. Phys. Lett.*, **207** (1993) 339.

111. R. Malhotra, S. Kumar and A. Satyam, *J. Chem. Soc., Chem. Commun.*, (1994) 1339.

112. J. Deng, C. Mou and C. Han, *J. Phys. Chem.*, **99** (1995) 14907; *Fullerene Sci. & Technol.*, **5** (1997) 1325.

113. J. Deng *et al.*, *J. Phys. Chem.*, **97** (1993) 11575.

114. T. Nogami, M. Tsuda, T. Ishida, S. Kurono and M. Ohashi, *Fullerene Sci. & Technol.*, **1** (1993) 275.

115. A. L. Balch, D. A. Costa, B. C. Noll and M. M. Olmstead, *J. Am. Chem. Soc.*, **117** (1995) 8926.

116. R. Taylor, N. J. Tower, M. Barrow and T. A. Drewello, unpublished work.

117. R. Taylor *et al.*, *J. Chem. Soc., Perkin Trans. 2*, (1995) 181.

118. S. W. McElvaney and C. L. Holliman, *Recent Advances in the Chem. and Phys. of Fullerenes*, ed. K. Kadish and R. S. Ruoff, **3** (1996) 121.

119. T. Hamano *et al.*, *Chem Commun.*, (1997) 21.
120. A. L. Balch, D. A. Costa, B. C. Noll and M. M. Olmstead, *Inorg. Chem.*, **35** (1996) 458.
121. J. Deng, C. Mou and C. Han, *Chem. Phys. Lett.*, **256** (1996) 96.
122. T. Akasaka, W. Ando, K. Kobayashi and S. Nagase, *Fullerene Sci. & Technol.*, **1** (1993) 339.
123. S. R. Wilson, Y. Wu, N. A. Kaprinidis, D. I. Schuster and C. J. Welch, *J. Org. Chem.*, **58** (1993) 6548; S. R. Wilson, N. Kaprininidis, Y. Wu and D. I. Schuster, **115** (1993) 8495; D. I. Schuster *et al.*, **188** (1996) 5639.
124. X. Zhang, A. Romero and C. S. Foote, *J. Am. Chem. Soc.*, **115** (1993) 11024; X. Zhang, A. Fan and C. S. Foote, *J. Org. Chem.*, **61** (1996) 5456.
125. X. Zhang and C. S. Foote, *J. Am. Chem. Soc.*, **117** (1995) 4271.
126. S. Yagamo, A. Takeuchi and E. Nakamura, *J. Am. Chem. Soc.*, **116** (1994) 1123.
127. G. Vassilikogannikis and M. Orfanopoulos, *J. Am. Chem. Soc.*, **119** (1997) 7394.
128. L. Shu *et al.*, *Chem. Commun.*, (1997) 79; B. F. O'Donovan, P. B. Hitchcock, M. F. Meidine, H. W. Kroto, R. Taylor and D. R. M. Walton, *Chem. Commun.*, (1997) 81.
129. K. Liou and C. Cheng, *J. Chem. Soc., Chem. Commun.*, (1995) 2473.
130. S. Wu, L. Shu and K. Fan, *Tetrahedron Lett.*, **356** (1994) 919; K. Komatsu, Y. Murata, N. Sugita and T. S. M. Wan, *Chem. Lett.*, (1994) 635.
131. H. Tokuyama, H. Isobe and E. Nakamura, *J. Chem. Soc., Chem. Commun.*, (1994) 2753.
132. X. Zhang and C. S. Foote, *J. Org. Chem.*, **59** (1994) 5235.
133. G. Mehta and M. J. Viswanath, *Tetrahedron Lett.*, **36** (1995) 5631.
134. T. Kusakuwa, A. Shike and W. Ando, *Tetrahedron*, **52** (1996) 4995.
135. M. Prato, M. Maggini, G. Scorrano and V. Lucchini, *J. Org. Chem.*, **58** (1993) 3613.
136. S. H. Hoke *et al.*, *J. Org. Chem.*, **57** (1992) 5069.
137. M. Tsuda, T. Ishida, T. Nogami, S. Kurono and M. Ohashi, *Chem. Lett.*, (1992) 2333.
138. A. D. Darwish, A. K. Abdul-Sada, G. J. Langley, H. W. Kroto, R. Taylor and D. M. Walton, *J. Chem. Soc., Chem. Commun.*, (1994) 213;

A. D. Darwish, A. G. Avent, R. Taylor and D. R. M. Walton, *J. Chem. Soc., Perkin Trans. 2*, (1996) 2079.

139. M. Meier, G. Wang, R. C. Haddon, C. P. Brock, A. A. Lloyd and J. P. Selegue, *J. Am. Chem. Soc.*, **120** (1998) 2337.

140. S. H. Hoke, J. Molstad, S. Yang, D. Carlson and B. Kahr, *J. Org. Chem.*, **59** (1994) 3230.

141. A. D. Darwish and R. Taylor, unpublished work.

142. Z. Slanina, S. Lee and R. Taylor, *Synthetic Metals*, **77** (1996) 51.

143. L. Shiu *et al.*, *J. Chem. Soc., Chem. Commun.*, (1994) 647.

144. T. Suzuki, Y. Maruyama, T. Akasaka, W. Ando, K. Kobayshi and S. Nagase, *J. Am. Chem. Soc.*, **116** (1994) 1359.

145. S. Yamago and E. Nakamura, *Chem. Lett.*, (1996) 395; see also S. Yamago, H. Tokuyama, E. Nakamura, M. Prato and F. Wudl, *J. Org. Chem.*, **58** (1993) 4796.

146. E. Nakamura, H. Isobe, H. Tokuyama and M. Sawamura, *Chem. Commun.*, (1996) 1747.

147. E. Beer, M. Feuerer, A. Knorr, A. Mirlach and J. Daub, *Angew. Chem. Intl. Edn. Engl.*, **33** (1994) 1087.

148. M. Maggini, G. Scorrano and M. Prato, *J. Am. Chem. Soc.*, **115** (1993) 9798; M. Prato, M. Maggini, C. Giacometti, G. Scorrano, G. Sandonà and G. Farnia, *Tetrahedron*, **52** (1996) 5221.

149. M. Maggini, A. Karlsson, G. Scorrano, G. Sandonà, G. Farnia and M. Prato, *J. Chem. Soc., Chem. Commun.*, (1994) 589; N. Martin, I. Pérez, L. Sánchez and C. Seoane, *J. Org. Chem.*, **62** (1997) 5690.

150. N. Martin, L. Sanchez, C. Seoane, R. Andreu, J. Garin and J. Orduna, *Tetrahedron Lett.*, **37** (1996) 5979.

151. M. Maggini, A. Donô, G. Scorrano and M. Prato, *J. Chem. Soc., Chem. Commun.*, (1995) 845.

152. T. Drovetskaya, C. A. Reed and P. D. W. Boyd, *Tetrahedron Lett.*, **36** (1995) 7971; Y. Sun, T. Drovetskaya, R. D. Bolskar, R. Bau, P. D. W. Boyd and C. A. Reed, *J. Org. Chem.*, **62** (1997) 3642; H. Imahori and Y. Sakata, *Chem. Lett.*, (1996) 199.

153. X. Shi, W. B. Caldwell, K. Chen and C. A. Mirkin, *J. Am. Chem. Soc.*, **116** (1994) 11598.

154. F. Novello *et al.*, *Chem. Commun.*, (1996) 903.

155. T. De Ros, M. Prato, F. Novello, M. Maggini and E. Banfi, *J. Org. Chem.*, **61** (1996) 9070.

156. R. Signorini *et al.*, *Chem. Commun.*, (1996) 1891.

157. S. R. Wilson, Y. Wang, J. Cao and X. Tan, *Tetrahedron Lett.*, **37** (1996) 775; L. Shu, G. Wang, S. Wu, H. Wu and X. Lao, *Tetrahedron Lett.*, **36** (1995) 3871.

158. M. Iyoda *et al.*, *Chem. Lett.*, (1997) 63.

159. L. Gan *et al.*, *J. Org. Chem.*, **61** (1996) 1954.

160. A. Komori, M. Kubota, T. Ishida, H. Niwa and T. Nogami, *Tetrahedron Lett.*, **37** (1996) 4031.

161. Q. Li, D. I. Schuster and S. R. Wilson, *J. Org. Chem.*, **61** (1996) 4764.

162. W. Duczek and H. Niclas, *Tetrahedron Lett.*, **36** (1995) 2457.

163. M. Maggini *et al.*, *J. Chem. Soc., Chem. Commun.*, (1994) 305.

164. A. Bianco *et al.*, *J. Am. Chem. Soc.*, **119** (1997) 7550.

165. M. Maggini, A. Karlsson, L. Pasimeni, G. Scorrano, M. Prato and L. Valli, *Tetrahedron Lett.*, **35** (1994) 2985.

166. J. Averdung *et al.*, *Chem. Ber.*, **127** (1994) 787.

167. A. A. Ovcharenko, V. A. Chertkov, A. V. Karchava and M. A. Yurovskaya, *Tetrahedron Lett.*, **39** (1997) 6933.

168. W. Duczek and H. Niclas, *Tetrahedron Lett.*, **36** (1995) 2457.

169. S. Muthu, P. Maruthamuthu, R. Ragunathan, P. R. Vasudeva Rao and C. K. Matthews, *Tetrahedron Lett.*, **35** (1994) 1763; Y. Matsubara, H. Tada, S. Nagase and Z. Yoshida, *J. Org. Chem.*, **60** (1995) 5372.

170. M. S. Meier and M. Poplawska, *J. Org. Chem.*, **58** (1993) 4524, *Tetrahedron*, **52** (1996) 5043.

171. H. Irngartinger, C. Köhler, U. Huber-Patz and W. Krätschmer, *Chem. Ber.*, **127** (1994) 581; H. Irngartinger, A. Weber, T. Escher, *Liebigs Ann. Chem.*, (1996) 1845.

172. T. Da Ros, M. Prato, F. Novello, M. Maggini, M. De Amici and C. De Micheli, *Chem. Commun.*, (1997) 59.

173. S. Eguchi *et al.*, *Fullerene Sci. & Technol.*, **4** (1996) 303.

174. H. Irngartinger and A Weber, *Tetrahedron Lett.*, **38** (1997) 2075.

175. M. S. Meier, M. Poplawska, A. L. Compton, J. P. Shaw, J. P. Selegue and T. F. Guarr, *J. Am. Chem. Soc.*, **116** (1994) 7044; H. Irngartinger, C. Köhler, G. Baum and D. Fenske, *Leibigs Ann. Chem.*, (1996) 1609.

176. J. Averdung, J. Mattay, D. Jacobi and W. Abraham, *Tetrahedron*, **51** (1995) 2543.

177. M. R. Banks *et al.*, *Tetrahedron Lett.*, (1994) 9067.

178. L. Shiu *et al.*, *J. Chem. Soc.*, *Perkin Trans. 1*, (1994) 3355.

179. D. Brizzolara, J. T. Ahlemann, H. W. Roesky and K. Keller, *Bull. Soc. Chim Fr.*, **130** (1993) 745.

180. M. Ohno, A. Yashiro and S. Eguchi, *Chem. Commun.*, (1996) 291.

181. N. Jagerovic, J. Elguero and J. Aubagnac, *J. Chem. Soc.*, *Perkin Trans. 1*, (1996) 499.

182. C. K. F. Shen, K. Chien, T. Liu, T. Lin, G. Her and T. Luh, *Tetrahedron Lett.*, **36** (1995) 5383.

183. T. Akasaka, W. Ando, K. Kobayashi and S. Nagase, *J. Am. Chem. Soc.*, **115** (1993) 10366.

184. T. Akasaka, E. Misuhida, W. Ando, K. Kobayashi and S. Nagase, *J. Am. Chem. Soc.*, **116** (1994) 2527.

185. M. F. Meidine *et al.*, *J. Chem. Soc.*, *Chem. Commun.*, (1993) 1342.

186. P. Belik, A. Gügel, J. Spickermann and K. Müllen, *Angew. Chem. Intl. Edn. Engl.*, **32** (1993) 78.

187. P. Belik, A. Gügel, A. Kraus, M. Walter and K. Müllen, *J. Org. Chem.*, **60** (1995) 3307.

188. M. Walter, A. Gügel, J. Spickermann, P.Belik, A. Kraus and K. Müllen, *Fullerene Sci. & Technol.*, **4** (1996) 101.

189. F. Diederich, U. Jonas, V. Gramlich, A. Hermann, H. Ringsdorf and C. Thilgen, *Helv. Chim. Acta*, **76** (1993) 2445.

190. Y. Nakamura, T. Minowa, S. Tobita, H. Shizuka and J. Nishimura, *J. Chem. Soc.*, *Perkin Trans. 2*, (1995) 2351.

191. W. Bidell, R. E. Douthwaite, M. L. H. Green, A. H. H. Stephens and J. F. C. Turner, *J. Chem. Soc.*, *Chem. Commun.*, (1994) 1641.

192. A. Hermann, F. Diedereich, C. Thilgen, H. ter Meer and W. H. Müller, *Helv. Chim. Acta*, **77** (1994) 1689; P. Seiler, A. Hermann and F. Diederich, *Helv. Chim. Acta*, **78** (1995) 344.

193. M. Taki *et al.*, *J. Am. Chem. Soc.*, **119** (1997) 926.

194. L. A. Paquette and R. J. Graham, *J. Org. Chem.*, **60** (1995) 2958; L. A. Paquette and W. E. Trego, *Chem. Commun.*, (1996) 419.

195. U. M. Fernández-Paniagua, B. M. Illecas, N. Martin and C. Seoane, *J. Chem. Soc.*, *Perkin Trans. 1*, (1996) 1077.

196. M. Ohno, N. Koide, H. Sato, S. Eguchi, *Tetrahedron*, **53** (1997) 9075.

197. M. Walter, A. Gügel, J. Spickermann, P. Belik, A. Kraus and K. Müllen, *Fullerene Sci. & Technol.*, **4** (1996) 101.

198. A. C. Tomé, R. F. Enes, J. A. S. Cavaleiro and J. Elguero, *Tetrahedron Lett.*, **38** (1997) 2557.

199. J. Llacey, M. Mas, E. Molins, J. Veciana, D. Powell and C. Rovira, *Chem. Commun.*, (1997) 659; C. Boulle *et al.*, *Tetrahedron Lett.*, **38** (1997) 81, 3909.

200. A. Gügel *et al.*, *Tetrahedron*, **52** (1996) 5007.

201. B. M. Illescas, N. Martin, C. Seoane, P. de la Cruz, F. Langa and F. Wudl, *Tetrahedron Lett.*, **36** (1995) 8307; B. M. Illescas *et al.*, *J. Org. Chem.*, **62** (1997) 7585.

202. Y. Rubin, S. Khan, D. I. Freedberg and C. Yeretzian, *J. Am. Chem. Soc.*, **115** (1993) 344.

203. H. Tomioka, M. Ichihashi and K. Yamamoto, *Tetrahedron Lett.*, **36** (1995) 5371.

204. M. Ohno, T. Azuma and S. Eguchi, *Chem. Lett.*, (1993) 1833; P. de la Cruz, A. de la Hoz, F. Langa, B. Illescas and N. Martin, *Tetrahedron*, **53** (1997) 2599.

205. M. Prato *et al.*, *J. Am. Chem. Soc.*, **115** (1993) 1594; R. Malhotra *et al.*, *Mat. Res. Soc. Symp. Proc.*, **247** (1992) 301; W. Bidell, R. E. Douthwaite, M. L. H. Green, A. H. H. Stephens and J. F. C. Turner, *J. Chem. Soc.*, *Chem. Commun.*, (1994) 1641.

206. A. Puplovskis, J. Kacens and O. Neilands, *Tetrahedron Lett.*, **38** (1997) 285.

207. A. Gügel, A. Kraus, J. Spickermann, P. Belik and K. Müllen, *Angew. Chem. Intl. Edn. Engl.*, **33** (1994) 559.

208. M. Iyoda, F. Sultana, S. Sasaki and M. Yoshida, *J. Chem. Soc.*, *Chem. Commun.*, (1994) 1929; H. Tomioka and K. Yamamoto, *J. Chem. Soc.*, *Chem. Commun.*, (1995) 1961; M. Iyoda, F. Sultana, S. Sasaki and H. Butenaschön, *Tetrahedron Lett.*, **36** (1995) 579; M. Iyoda, S. Sasaki, F. Sultana, M. Yoshida, Y. Kuwatani and S. Nagase, *Tetrahedron Lett.*, **37** (1996) 7987; A. Herrera, R. Martinez, B. González, N. Martin and C. Soanne, *Tetrahedron Lett.*, **38** (1997) 4873; M. Ohno, S. Kojima, Y. Shirakawa and S. Eguchi, *Tetrahedron Lett.*, **36** (1995) 6899.

209. B. Kräutler and M. Puchberger, *Helv. Chim. Acta*, **76** (1993) 1626; J. Maynollo and B. Kräutler, *Fullerene Sci. & Technol.*, **4** (1996) 213; *Tetrahedron*, **52** (1996) 5033.

210. J. Mestro, M. Duran and M. Sola, *J. Phys. Chem.*, **100** (1996) 7449.

211. J. Pola and R. Taylor, unpublished work.

212. M. Ohno, T. Azuma, S. Kojima, Y. Shirakawa and S. Eguchi, *Tetrahedron*, **52** (1996) 4983.

213. S. R. Wilson and Q. Li, *Tetrahedron Lett.*, **34** (1993) 8043.

214. Y. An, J. L. Anderson and Y. Rubin, *J. Org. Chem.*, **58** (1993) 4799; Y. An, C. B. Chen, J. L. Anderson, D. S. Sigman, C. S. Foote and Y. Rubin, *Tetrahedron*, **52** (1996) 5179.

215. M. Arce, A. L. Viado, Y. An, S. I. Khan and Y. Rubin, *J. Am. Chem. Soc.*, **118** (1996) 3775.

216. L. Isaacs, R. F. Haldimann and F. Diederich, *Angew. Chem. Intl. Edn. Engl.*, **33** (1994) 2339.

217. T. G. Linssen, K. Dürr, M. Hanack and A. Hirsch, *J. Chem. Soc., Chem. Commun.*, (1995) 103; *Chem. Ber./Recueil*, **130** (1997) 1375.

218. G. Torres-Garcia and J. Mattay, *Tetrahedron*, **52** (1996) 5421; G. Torres-Garcia, H. Luftmann, C. Wolff and J. Mattay, *J. Org. Chem.*, **62** (1997) 2752.

219. J. Liu, A. Mori, N. Kato and H. Takeshita, *Chem. Lett.*, (1993) 1697; *Tetrahedron Lett.*, **35** (1994) 6305; *Fullerene Sci. & Technol.*, **3** (1995) 45.

220. J. Liu, N. Kato, A. Mori, H. Takeshita and R. Isobe, *Bull Chem. Soc. Jpn.*, **67** (1994) 1507.

221. H. Takeshita, J. Liu, N. Kato, A. Mori and R. Isobe, *Chem. Lett.*, (1995) 377.

222. M. Ohkita, K. Ishigami and T. Tsuji, *J. Chem. Soc., Chem. Commun.*, (1995) 1769.

223. M. Ohno, S. Kojima, Y. Shirakawa and S. Eguchi, *Tetrahedron Lett.*, **37** (1996) 9211; M. Ohno, S. Kojima and S. Eguchi, *J. Chem. Soc., Chem. Commun.*, (1995) 565.

224. K. Liou and C. Cheng, *J. Chem. Soc., Chem. Commun.*, (1995) 1603.

225. J. A. Schleuter *et al.*, *J. Chem. Soc., Chem. Commun.*, (1993) 972; K. Komatsu, Y. Murata, K. Sugita, K. Takeuchi and T. S. M. Wan, *Tetrahedron Lett.*, **34** (1993) 8473.

226. B. Kräutler *et al.*, *Angew. Chem. Intl. Edn. Engl.*, **35** (1996) 1204.

227. I. Lamparth, C. Maichle-Mössmer and A. Hirsch, *Angew. Chem. Intl. Edn. Engl.*, **34** (1995) 1607.

228. A. G. Avent, P. R. Birkett, A. D. Darwish, H. W. Kroto, R. Taylor and D. M. Walton, *Fullerene Sci. & Technol.*, **5** (1997) 643.

229. L. S. K. Pang and M. A. Wilson, *J. Phys. Chem.*, **97** (1993) 6761; V. M. Rotello *et al.*, *Tetrahedron Lett.*, **34** (1993) 1561.

230. M. F. Meidine, A. G. Avent, A. D. Darwish, O. Ohashi, R. Taylor and D. R. M. Walton, *J. Chem. Soc., Perkin Trans. 2*, (1994) 1189.

231. M. F. Meidine *et al.*, *J. Chem. Soc., Perkin Trans. 2*, (1994) 2125.

232. S. Ballenweg, R. Gleiter and W. Krätschmer, *J. Chem. Soc., Chem. Commun.*, (1994) 2269.

233. R. N. Warrener, G. M. Elsey and M. A. Houghton, *J. Chem. Soc., Chem. Commun.*, (1995) 1417.

234. A. Mehte, A. Ulug and B. Ulug, *Fullerene Sci. & Technol.*, **4** (1996) 457.

235. B. Nie, K. Hasan, M. D. Greaves and V. Rotello, *Tetrahedron Lett.*, **36** (1995) 3617.

236. M. R. Banks *et al.*, *J. Chem. Soc., Chem. Commun.*, (1995) 1171.

10

Oxidation and the Formation of Radical Cations and Cations

Three aspects are covered in this chapter: reaction with oxygen (and oxygen-containing species), and formal oxidation either by the removal of one electron from a π-bond on the cage (giving a radical cation), or through heterolytic cleavage of a σ-bond to an addend (giving a cation). The former process takes place readily but the latter two processes are more difficult to achieve.

10.1 Addition of Oxygen

Two processes are considered here, namely the addition of oxygen to give an epoxide, covered in more detail in Sec. 9.1.3, and oxidative degradation of the cage. These processes are an inherent problem when dealing with fullerenes, and are exacerbated by the ability of [60]fullerene to absorb up to 4% O_2 in the lattice;[1] the enhanced degradation in the presence of oxygen was observed at an early stage.[2,3]

Oxidative degradation of the cage can be brought about by oxygen[4] (which is converted to singlet oxygen by the fullerene),[5] ozone[2,6] and peracids. In ozone oxidation of fullerenes, [70]fullerene is the least reactive;[7] cage degradation, which accompanies epoxide formation by 3-chloroperbenzoic acid is more severe for [70]fullerene than for [60]fullerene.[8] Very thin films of fullerenes, deposited on glass are oxidatively degraded within a few days, and irradiation of hexane solutions of fullerenes by a medium pressure UV lamp causes complete degradation within 24 h; degradation of benzene solutions is a little slower.[2] The nature of the products from these degradations have not been fully characterised, but IR spectroscopy of these and other oxidations indicates the presence variously of fullerene oxides, carbonyl-containing derivatives (acids, anhydrides etc.), fullerenols (which are water soluble), and ultimately CO and CO_2.[2,4,6,9] The reaction with oxygen under irradiation is accompanied by intensification of the weak ESR signal that [60]fullerene produces, indicating that this signal (which should be absent) arises from oxygenated impurities.[10] In the absence of light, oxidative degradation of [60]fullerene has been shown by a combination of X-ray absorption-, photoelectron-, and diffuse reflectance IR spectroscopy to commence at around 470 K, and leads to complete degradation of molecules in the surface region at 570 K.[11]

The UV-catalysed oxidative degradation of [60]fullerene was discovered by the writer during column chromatography on neutral alumina,[2] and is so rapid that the alumina must play a catalytic role. (This caused considerable problems in preparing, for a BBC *Tomorrow's World* programme in 1991, alumina columns showing the separation of fullerenes.) At rehearsal, the [60]fullerene layer turned from magenta to brown in only a few minutes under the bright lights of the television studio, so it was necessary to keep the columns wrapped in aluminium foil until just before the live transmission.) Titanium oxide also has a catalytic effect on the oxidative degradation of [70]fullerene,[12] and it is possible that very fine alumina particles contribute to the spectacular oxidative degradation that can be observed by standing a benzene solution of [60]fullerene over water in a closed vessel near to a window. Over time, the upper magenta layer becomes colourless, whilst the lower layer becomes orange-brown. However, the rate at which this occurs is very dependent upon the fullerene sample, so that some other species present as an impurity must catalyse the process. [70]Fullerene does not show this behaviour.

Both chemical and electrochemical oxidation of [60]fullerene in either $HClO_4$, H_2SO_4, or HNO_3 gives orange to brown products, which have not been fully characterised,[13] though the product from the nitric acid oxidation has the characteristics (IR, solubility) of a fullerenol (see Sec. 8.2).

10.2 Formation of Radical Cations

Both [60]- and [70]fullerenes have low-lying HOMOS, so that electrochemical oxidation is quite difficult, but has been achieved. The chemically reversible one-electron oxidation waves (giving radical cations with lifetimes > 30 s) occur at 1.26 and 1.20 v, respectively; for [70]fullerene a second one-electron oxidation is observed at 1.75 v, resulting in the formation of a dication.[14] As in the case of dianion formation, the intermediate structure may in principle be a diradical cation, but intramolecular coupling of the radicals leads to double bond formation and a dication. Oxidation of [76]fullerene is easier, the electrochemical oxidation potential being 0.81 V.[15]

[60]Fullerene is oxidised to $C_{60}^{\bullet+}$ (detected by an ESR signal at 2.0027 g) by the Fe^{3+} in a Fe^{3+}-exchanged zeolite. The ESR signal disappeared on addition of perylene, being replaced by a new one due to the perylene radical cation, as a result of electron transfer from perylene to the fullerene,[16] and follows from the one-electron oxidation potential of perylene (1.04 V) being lower than the one-electron reduction potential of $C_{60}^{\bullet+}$ (1.57 V).

The [60]fullerene radical cation (showing the characteristic[17] near-IR bands at 960 and 980 nm) is also formed by irradiation of solutions of the fullerene in 1,2-dichloroethane, conditions which produce highly oxidising 1,2-DCE radical cations.[18] The radical cation is also produced by irradiating solutions of [60]fullerene in the presence of 9,10-dicyanoanthracene as electron-acceptor sensitiser. Its presence can be detected by trapping with alcohols or alkanes, processes which proceed by initial hydrogen abstraction in each case;[19,20] in particular, reaction with alcohols RCH_2OH results in addition of H and RCHOH, rather than H and OCH_2R, conjectured from earlier mass spectrometry results (*cf.* Ref. 19).

$C_{60}^{\bullet+}$ is believed to be the intermediate in the reaction of [60]fullerene with peroxides $(RCO_2)_2O$, $(R = CF_3$ or $C_3F_7)$ which results in 1,4-addition of HO and R.[21]

The relative reactivities $(C_{60}^{\bullet+}/C_{70}^{\bullet+})$ in the gas phase towards 1,3-cyclopentadiene and 1,3-cyclohexadiene are *ca.* 5 in each case,[22] which parallels the greater reactivity of [60]fullerene over that of [70]fullerene towards cycloaddition (Sec. 9.4.3).

In the above reactions, formation of the radical cations has been inferred by spectroscopic means, but isolation of them is altogether more demanding. Any oxidising reagent is likely to be associated with a strong nucleophile, which will react and thus destroy the produced species. This has been overcome, at least for the formation of $C_{76}^{\bullet+}$, by using an extremely unreactive nucleophilic counterion, $CB_{11}H_6Br_6^-$, the overall oxidant being $[Ar_3N^{\bullet-}]\,[CB_{11}H_6Br_6^-]$. The

IR spectrum of $C_{76}^{\bullet+}$ is identical to that of the parent fullerene except that the band at 1440 cm^{-1} is shifted to 1456 cm^{-1}; an ESR spectrum is observed with $g = 2.0030$ (O.5 G linewidth).[15]

10.3 Formation of Cations

As noted in Sec. 10.2, formation of a di(radical cation) can result in conversion to a dication. These species (from both [60]- and [70]fullerenes) are produced by 50 eV electron impact on the fullerene in argon gas. They act as polymerisation initiators of buta-1,3-diene, with equal effectiveness; by contrast the radical cations are unreactive. The cations are assumed to be on opposite sides of the molecules (to minimise energy of repulsion) and hence that the fullerene is incorporated into the polymer chain (which grows on either side).[23]

The only other way to form a cation is by cleavage of a group from an sp^3-hybridised carbon of the cage. This process was thought earlier unlikely to be involved in nucleophilic substitution of halogenofullerenes (giving an S$_N$1-type process) and since an S$_N$2 process is sterically precluded, a novel elimination process (or alternatively an addition-elimination one) seemed probable. The latter is probably involved when no Lewis acid catalyst is present. However, it has now been shown (by ^{13}C NMR) that in the presence of AlCl$_3$, a free cation is formed from C$_{60}$Ph$_5$Cl (Fig. 8.6, X = Cl) but since loss of the halogen would produce an anti-aromatic (4π) pentagonal ring, the adjacent phenyl ring carries out a synchronous 1,2-shift, resulting in the formation of the rearranged cation and hence, in the presence of moisture, of an alcohol (see Fig. 8.7).[24] Formation of the cation makes the fullerene cage less aromatic, consequently the ^{3}He NMR spectrum (see Sec. 14.4) of a sample of i^3He C$_{60}$Ph$_5$Cl shows a downfield shift of 2 ppm on formation of the cation.[25]

References

1. T. Arai, Y. Murakami, H. Suematsu, K. Kikuchi Y. Achiba and I. Ikemoto, *Solid State Commun.*, **84** (1992) 827.
2. R. Taylor *et al.*, *Nature*, **357** (1991) 277.
3. I. Gilmour, J. P. Hare, H. W. Kroto and R. Taylor, *Lunar Planet Sci. Conf. XXII*, (1991) 445; M. Gevaert and P. M. Kamat, *J. Chem. Soc., Chem.*

Commun., (1992) 1470; H. S. Chen, A. R. Kortan, R. C. Haddon and D. A. Fleming, *J. Phys. Chem.*, **96** (1992) 1016.

4. C. Taliani *et al.*, *J. Chem. Soc.*, *Chem. Commun.*, (1993) 220.
5. J. W. Arbogast *et al.*, *J. Phys. Chem.*, **95** (1991) 11; J. W. Arbogast and C. S. Foote, *J. Am. Chem. Soc.*, **113** (1991) 8886.
6. R. Mahlhotra, S. Kumar and A. Satyam, *J. Chem. Soc.*, *Chem. Commun.*, (1994) 1339.
7. D. Heyman and L. P. F. Chibante, *Recl. Trav. Chim. Pays-Bas*, **112** (1993) 531, 639.
8. M. Al-Jafari, T. Drewello, N. J. Tower and R. Taylor, unpublished work.
9. G. H. Kroll *et al.*, *Chem. Phys. Lett.*, **181** (1991) 112; A. Datta, R. Y. Kelkar, P. Borojerdian, S. K. Kulkarni and M. Datta, *Bull. Chem. Soc. Jpn.*, **67** (1994) 1517; J. M. Wood, B. Kahr, S. H. Hoke, H. L. Dejarme, R. G. Cooks and D. Ben Amotz, *J. Am. Chem. Soc.*, **113** (1991) 5907; K. M. Creegan *et al.*, *J. Am. Chem. Soc.*, **114** (1992) 1103.
10. S. Kawata *et al.*, *Chem. Lett.*, (1992) 1659.
11. M. Wohlers, H. Werner, D. herein, T. Schedel-Niedrig, A. Bauer and R. Schlögl, *Synthetic Metals.*, **77** (1996) 299.
12. M. Gevaert and P. V. Kamat, *J. Chem. Soc.*, *Chem. Commun.*, (1992) 1470.
13. P. Scharf *et al.*, *Carbon*, **32** (1994) 709.
14. D. Dubois, K. M. Kadish, S. Flanagan and L. J. Wilson, *J. Am. Chem. Soc.*, **113** (1991) 7773; Q. Xie, F. Arias and L. Echegoyen, *J. Am. Chem. Soc.*, **115** (1993) 9818.
15. R. D. Bolskar, R. S. Mathur and C. A. Reed, *J. Am. Chem. Soc.*, **118** (1996) 13093.
16. S. Fukuzumi, T. Suenobu, T. Urano and K. Tanaka, *Chem. Lett.*, (1997) 875.
17. S. Nonell, J. W. Arbogast and C. S. Foote, *J. Phys. Chem.*, **96** (1992) 4169.
18. D. M. Guldi, H. Hungerbühler, E. Janata and K. Asmus, *J. Chem. Soc.*, *Chem. Commun.*, (1993) 84.
19. G. Lem, D. I. Schuster, S. H. Courtney, Q. Li and S. R. Wilson, *J. Am. Chem. Soc.*, **117** (1995) 554.
20. C. Siedschlag, H. Luftmann, C. Wolff and J. Mattay, *Tetrahedron*, **53** (1997) 3587.

21. M. Yoshida, Y. Morinaga, M. Iyoda, K. Kikuchi, I. Ikemoto and Y. Achiba, *Tetrahedron Lett.*, **34** (1993) 7629.
22. H. Becker, G. Javahery, S. Petri and D. K. Bohme, *J. Phys. Chem.*, **98** (1994) 5591.
23. J. Wang, G. Javahery, S. Petri and D. K. Bohme, *J. Am. Chem. Soc.*, **114** (1992) 9665.
24. P. R. Birkett, A. G. Avent and R. Taylor, unpublished work.
25. P. R. Birkett, A. Khong, M. Saunders and R. Taylor, unpublished work.

11

Inorganic and Organometallic Derivatives of Fullerenes

11.1 Introduction

A notable feature of these fullerene derivatives (which are generally highly coloured) is the ease with which a wide variety of them can be made (in high yield), and the ready isolation of crystalline derivatives which are amenable to examination by X-ray crystallography. This has been of great value in determining both the structures of the fullerene cages, and the location of addends, though by contrast none of the derivatives have yet found use in subsequent syntheses. The bonding in many cases is of a co-ordination nature, resulting in derivatives which tend to be of relatively low stability. There is also evidence from a number of studies that the metal groups migrate over the cage surface. Early anticipation that fullerenes would be aromatic led to the expectation of the formation of η^6-derivatives. However, the fullerenes are electron-deficient alkenes and thus η^2-derivatives are obtained instead, though in one case where localisation of two double bonds in a pentagon occurs, a η^5-derivative has been prepared.

The formation of some organometallic fullerene derivatives have been described elsewhere in this book, and are not therefore described further here. Examples are the ball and chain structures noted in Sec. 12.7, and the anion complexes formed by electron transfer from organometallic donors (Sec. 5.3).

11.2 Osmylation

Osmylation is a typical reaction of osmium tetroxide with alkenes, giving an osmate ester which can be hydrolytically cleaved to give *cis* 1,2-diols. If the

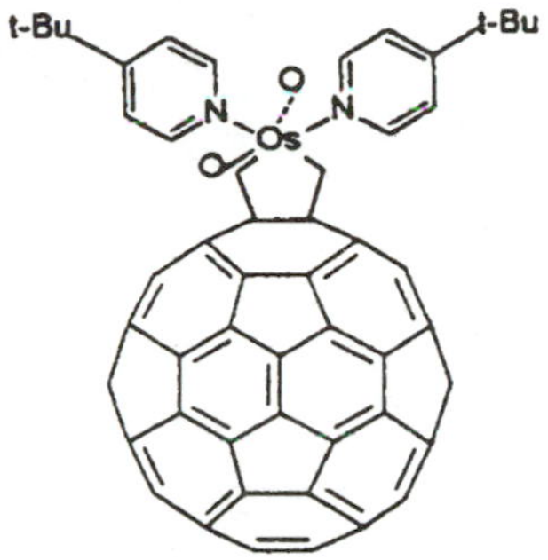

11.1

reaction is carried out in the presence of pyridine, stable osmate ester adducts can be isolated. Hawkins and coworkers used the reaction very effectively in the determination of fullerene structures, the most notable being **11.1**[1] which provided the first crystallographic confirmation of the structure of [60]fullerene, (determined earlier by ^{13}C NMR),[2] and showed that addition occurred across a 6,6-bond. This derivative utilised 4-*t*-butylpyridine as the ligand because it improved the solubility and crystal quality compared to the product obtained by the use of pyridine.

Bis-osmylation yields only five of the possible eight regioisomers, due evidently to the size of the addend. Two of the isomers have C_2 symmetry and two have C_s symmetry. A combination of 1- and 2-D ^{13}C NMR and the HPLC retention times shows two isomers to be *e* and *trans*-3, the remaining three isomers being very probably *trans*-2, *trans*-4 and *trans*-1 (see Fig. 2.10). The latter isomer elutes the most rapidly (consistent with it having the non-polar D_{2h} structure) and it was also obtained in lowest yield,[3,4] a consequence of it being statistically four times less likely to be formed compared to any of the other isomers which have four equivalent double bonds available for the second addition. This lower yield is the reason for the failure to determine the symmetry.

The isomers of C_2 symmetry are chiral and are resolvable by using Sharpless cinchona alkaloids as the ligands instead of pyridine in the addition process. Subsequent exchange of these ligands for pyridine yields the resolved osmate esters. The alkaloids have a considerable effect on the relative isomer ratios, due probably to attractive interactions between the ligands and the cages rather than to steric effects.[4]

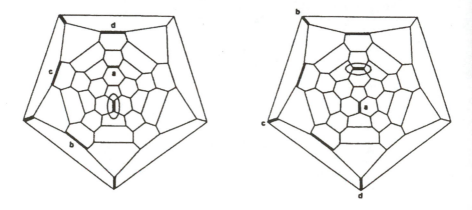

Figs. 11.1 and **11.2** Schlegel diagrams showing the feasible locations of the second addend on osmylation of 1,2- and 5,6-osmylated [70]fullerenes, respectively.

Osmylation of [70]fullerene gives addition across the 1,2- and 5,6-bonds in 2.1:1 ratio.[5] This parallels the behaviour of this molecule in all other additions (see Sec. 3.5), and contrasts with the reverse reactivity order predicted from the π-bond orders. The difference is generally ascribed to the greater curvature across the 1,2-bond, creating greater localised strain which is relieved upon addition. Further osmylation of each isomer gives six $C_{70}(OsO_4py_2)_2$ adducts from the 1,2-isomer and seven from the 5,6-isomer (eight are possible in each case), and four of the diadducts from each set appear to be common. Figs. 11.1 and 11.2 shows the sterically accessible di-adducts obtainable from each isomer, from which it can be seen that four of them, (the second addend sites are labelled a–d in each Figure) are the same in each case. (Note that these are assigned with the benefit of hindsight, and are not the locations given in the original analysis which are unlikely to be correct.) A similar pattern could be expected for other di-additions to [70]fullerene involving bulky reagents, but no comparable data are yet available.

One of the most elegant achievements in fullerene chemistry, has been the resolution of [76]fullerene (which is chiral) into its two enantiomers, by taking advantage of their different reactivities.[6] On reacting with osmium tetroxide and a chiral Sharpless ligand (in toluene at 0°C), [76]fullerene preferentially forms only one of the pseudoenantiomeric derivatives, and after removal from this of the osmate ligand (by $SnCl_2$ in pyridine), the recovered [76]fullerene has the opposite sign of rotation from that of the [76]fullerene which did not react. Furthermore, each isomer had mirror image CD spectra. The maximum specific rotation of [76]fullerene so derived is estimated to be $4000 \pm 400°$ which is comparable to that of the helicenes. The site of addition is believed, on the basis of maximum local curvature, to be across either the 2,3- or 1,6-bonds, which are also predicted to have amongst the highest π-densities in the molecule.[7] (See also Secs. 9.1.1.3 and 9.4.1.)

11.3 Fullerene Complexes of Iridium, Rhodium and Cobalt

Complexes formed with iridium compounds of the general formula $Ir(CO)Cl(PR_3)_2$ (Vaska's complex) have been used very successfully by Balch and coworkers for the structural characterisation of fullerenes and derivatives. Addition of the heavy iridium atom to the cage slows rotation thus reducing disorder and allowing the determination of single crystal X-ray structures. The complexes formed generally involve η^2-coordination to a 6,6-bond but the bonding is very weak, and the complexes dissociate in solution. They are obtained by crystallisation from benzene, some molecules of which are usually contained within the lattice and make π-π contact with the fullerene cages.

The reaction (with R = Ph) gave early confirmation that addition takes place across a 6,6-bond in [60]fullerene,[8] whilst addition to [70]fullerene giving **11.2**, $(\eta^2$-$C_{60})Ir(CO)Cl(PPh_3)_2$, provided the first crystallographic structure for that molecule, the pole-pole distance and the equatorial diameter being 7.90 Å and 6.82 Å, respectively.[9] This reaction provided the first evidence that addition across the 1,2-bond of [70]fullerene is preferred, despite having a slightly lower bond order than the 5,6-bond (Table 2.1). The greater 'space' available nearer the pole for the bulky addend appeared responsible for the discrepancy, which now however is attributed to the greater pyramidalisation of the 1- and 2-carbons compared to the rest. The complex $Ir(CO)Cl(Me_2PPh)_2$ adds to [70]fullerene in a similar fashion.[10]

The reaction also provided the first confirmation of the structure of [84]fullerene, addition occurring preferentially across the 32,53-bond of the D_{2d} isomer (see Fig. 2.9) to give **11.3**, $(\eta^2$-$C_{84})Ir(CO)Cl(PPh_3)_2$,[11] this being predicted by π-bond calculations to be the most reactive one in the two main isomers of [84]fullerene.[12]

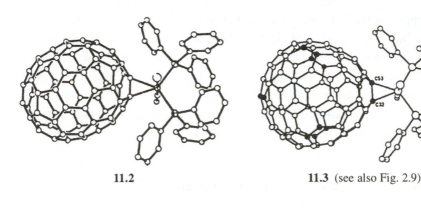

11.2   **11.3** (see also Fig. 2.9)

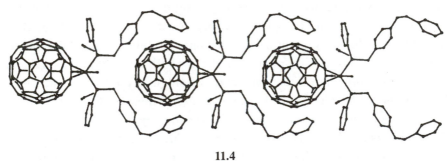

11.4

Replacement of the phenyl groups on phosphorus by other aromatics in Vaska's complex leads to some remarkable structures. For example, replacement of one phenyl group by -$CH_2C_6H_4OCH_2Ph$ produces the 'chain' structure shown in **11.4**, (η^2-C_{60})IrCOCl(Ph$_2$Phbob)$_2$, which differs from those described above, in having no benzene molecules in the crystal lattice. Instead the arms of the benzyloxybenzyl moiety form a 'cup' and thereby provides the π-π contact with the fullerene cages.[13] This structure is very specific for the size and shape of [60]fullerene, no comparable structure being obtainable from [70]fullerene.

The reactivity of Vaska's complex is increased by replacing the phenyl groups on phosphorus by electron-donating substituents, and this increased reactivity facilitates the formation of double-addition products. Thus each of the complexes Ir(CO)ClX_2, where X = PMe$_2$Ph, PEt$_3$ or PBu$^n_3$, added in large xs. to [60]fullerene result in addition across the 1,2- and 55,60-bonds.[14] Although such di-addition is least favoured statistically, it is dominant here most probably because of steric considerations. For X = PMe$_2$Ph, four different

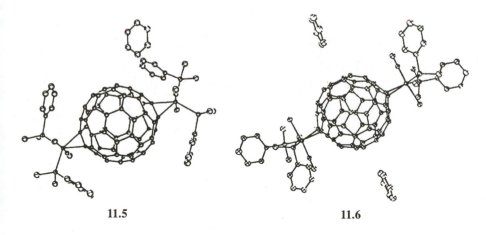

11.5   **11.6**

products were obtained, two of which have been characterised as shown in
11.5 and **11.6**, $(\eta^2\text{-}C_{60})Ir(CO)Cl(PMe_2Ph)_2$ conformers. In **11.5** the four phenyl
rings are folded back to make face-to-face contact with the cage surface, whereas
in the other, they face away from the cage. In this latter conformer however,
there are two benzene solvate molecules which are orientated perpendicular to
the cage, and in such a way that one of the hydrogens of benzene is 3.09 Å
above the electron-rich 18,36-bond that is common to two hexagonal rings.
Para di-addition is also obtained by reaction of [60]fullerene with the iridium
complex having $X = PEt_3$.[14]

Di-addition to [70]fullerene also occurs with Vaska's complex
($X = PMe_2Ph$). Given that monoaddition occurs across the 1,2-bond, then
reaction across the equivalent bonds at the opposite pole leads to three
possibilities (addition across either the (41,58-, 56,57-, or 67,68-bonds). In the
product that has been isolated and characterised, **11.7**, the second addition has
occurred across the 56,57-bond, but no reason for this preference is evident,[10]
though it should be noted that 56,57- and 67,68-addition have a 2:1 statistical
advantage over 41,58-addition (see also Sec. 9.1.1.5). The phenyl groups all
adopt a conformation so that they are *endo* to the cage as this maximises π-π
interactions, and the product should in principle be resolvable since it is chiral.

A novel di-adduct, **11.8**, is obtained by the reaction of $Ir_2(\mu\text{-}Cl)_2(\eta^4\text{-}cyclo\text{-}$
octadiene)$_2$ with [60]fullerene, and as in the reaction with Vaska-type
compounds, addition takes place without any ligand displacement.[15] Steric
considerations again favour *para* addition, but here iridium is doubly

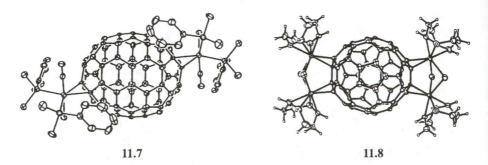

11.7   11.8

coordinated to the cage across the 1,2- and 3,4-bonds in the same hexagon. This is dictated by both steric considerations, and the fact that addition across one double bond in a hexagon localises the remaining two π-bonds, so that reaction occurs across one of them (see also Sec. 3.1). Three *para* isomers are possible, but only one (with addition across the 1,2,3,4- and 60,55,56,57-positions) has been isolated by crystallisation; two benzene molecules are incorporated in the lattice.

Iridium-fullerene complexes can also be formed by displacement of a ligand on iridium, by the fullerene. For example, reaction of [IrH(CO)(PPh$_3$)$_3$] with either [60]- or [70]fullerenes results in displacement of PPh$_3$ giving (η^2-C$_n$)IrHCO(PPh$_3$)$_2$ ($n = 60, 70$)[16], whilst the [60]fullerene compound can also be made by fullerene displacement of HCl from [IrH$_2$Cl(CO)(PPh$_3$)$_2$].[17] Displacement of cyclo-octene from the indenyl complex (η^5-C$_9$H$_7$)Ir(CO) (η^2-C$_8$H$_{14}$) by [60]fullerene gives (η^2-C$_{60}$)Ir(CO)(C$_9$H$_7$), the crystal structure of which (not yet determined) is believed to involve η^2-coordination of iridium to a 6,6-double bond of the fullerene.[18] Electrochemical reduction shows this to reversibly acquire two electrons each at a potential *ca.* 0.1 V more negative than the corresponding electron addition with [60]fullerene, and due to electron donation of the addend to the cage.

The formation of iridium complexes can be used to elucidate the structures of fullerene derivatives. For example on addition of Ir(CO)Cl(PPh$_3$)$_2$ to C$_{60}$O, iridium becomes coordinated to the 3,4-double bond in the ring in which oxygen bridges the 1,2-bond;[19] again the localisation of the π-electrons arising from the first addition is apparent. Note that here, the low steric requirement of the epoxide oxygen does not prevent local addition of the iridium complex. This is true also in the case of the iridium complexes of the principal isomers of

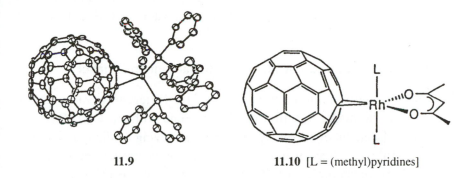

11.9   **11.10** [L = (methyl)pyridines]

$C_{60}O_2$, consisting of products resulting from a combination of 'S', 'T' and 'same ring' patterns (see Sec. 3.1) with the di-epoxide, in which the oxygens are across the 1,2- and 3,4-bonds in the same hexagonal ring.[20] This work provides confirmation, additional to that available from ^{13}C NMR, that the di-epoxide does not comprise an open-cage fullerene having two carbonyl groups in one ring (though such structures are probably present amongst the by-products of oxidations).

Rhodium complexes of both [60]- and [70]-fullerenes $(\eta^2\text{-}C_n)RhH(CO)$ $(PPh_3)_2$, $n = 60$, 70, involve addition across a 6,6-bond, and are green and dark brown powders, respectively,[16] and the structure of the [60]fullerene complex, $(\eta^2\text{-}C_{60})RhH(CO)(PPh_3)_2$, is **11.9**;[21,22] the complex is stable in solution, unlike the iridium analogue. A phosphine-free rhodium complex is **11.10**, $(\eta^2\text{-}C_{60})RhL_2acac$, L = (methyl)pyridines, in which rhodium is coordinated to acetylacetone and pyridine ligands.[23]

The ^{13}C NMR spectra of the rhodium complexes $(\eta^2\text{-}C_{60})RhH(CO)(PPh_3)_2$ and $(\eta^2\text{-}C_{60})Rh(NO)(PPh_3)_2$ change with increasing temperature, showing that fluxional process occur. Thus at $-90°C$, the spectra indicate a structure of C_s symmetry, but on warming to 50°C, this changes to a C_{2v} symmetry pattern showing that free rotation occurs about the cage-metal bond. At higher temperatures all resolution of the spectrum is lost, consistent with I_h symmetry, suggesting that the metal migrates over the cage surface, indicated by 2-D NMR to involve a succession of 1,3-shifts, as expected. Similar behaviour is found with $(\eta^2\text{-}C_{60})Ru(NO)(PPh_3)_2$ and $(\eta^2\text{-}C_{60})Co(NO)(PPh_3)_2$.[24] This observation of migration of organometallic groups confirms that similar behaviour is probable for platinum compounds, because intermediate di-, tri- etc. adducts are not on the pathway to a hexa-adduct (Sec. 11.4).

11.4 Fullerene Complexes of Platinum, Palladium and Nickel

Fullerene-platinum complexes were the first fullerene-organometallic derivatives to be made. Reaction of [60]fullerene with $(\eta^2\text{-}C_2H_4)Pt(PPh_3)_2$ results in replacement of ethene by the fullerene, giving an dihapto complex, $(\eta^2\text{-}C_{60})Pt(PPh_3)_2$.[25] This was contrary to the expectation (based on the contemporary presumption of [60]fullerene aromaticity) that hexahapto coordination to one ring would result, and served to emphasise that the fullerenes are polyalkenes rather than aromatics (see also Sec. 2.7.1). The electrochemical behaviour of this derivative showed that up to three electrons could be added reversibly, and at potentials *ca.* 0.3 V more negative relative to [60]fullerene itself because of the electron-donating effect of platinum ligand. Comparison with the data for the iridium ligands (Sec. 11.3) shows that the platinum ligand is more electron donating. The platinum derivative can also be made by displacement of HCl from $HPtCl(PtPh_3)_2$.[17]

In the analogous palladium complex, the C1-C2 bond is 1.447 Å compared to 1.502 Å for the platinum compound (and 1.39 Å for [60]fullerene itself). This indicates that coordination is less strong in the palladium compound. Another difference between the two is that the phenyl rings in the palladium complex are folded back to make stronger π-π contact with the cage.[26] The nickel compound $(\eta^2\text{-}C_{60})Ni(PEt_3)_2$ is also known, and notably the reduction potentials for the platinum, palladium and nickel compounds are virtually identical, despite the known differences in the electron-donating effect of the metals. The reduction potentials may therefore reflect the electron redistribution amongst the remaining double bonds following decoupling of one of them.[27]

If the smaller MEt_3 ligand (M = Pt, Pd) is reacted with [60]fullerene, up to six metallic groups co-ordinate to [60]fullerene in an octahedral pattern, as shown in e.g. **11.11** (the Et groups are omitted for clarity). This provided the first indication that formation of T_h symmetry structures (subsequently observed for example in cycloadditions) is an intrinsic property of this fullerene.[28] It arises because it creates increased aromaticity in eight of the hexagonal rings in the cage[29] (see also Sec. 3.2), and also minimises interactions between bulky

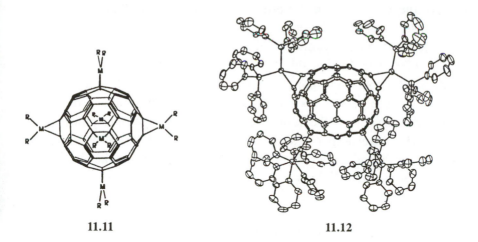

11.11   **11.12**

groups. The regiochemistries of the products of intermediate addition levels, $C_{60}[Pt(PEt_3)_2]_n$, $n = 2-5$, do not correspond to that of the hexa-adduct, indicating that migration of the platinum groups occurs on route to this latter. For example, two of the diadducts had the second addend in either of the *trans*-2, *trans*-3, or *trans*-4 positions, none of which lie on the required pathway (see Fig. 2.10). The platinum groups are rather labile, (and increasingly so as the addition level is increased, which follows from the decreasing electron withdrawal by the cage in this direction) and can, for example, be readily displaced by reaction with diphenylacetylene.

The larger size of [70]fullerene allows attachment of four of the larger $Pt(PPh_3)_2$ addends, and to the 1,2- and 31,32-bonds at one end of the molecule and the equivalent 41,58 and 65,66-bonds at the other (**11.12**),[30] locations which minimise steric interactions between the addends.

An alternative route to palladium and platinum complexes involves reaction of both phosphines and phosphites [PEt_3, PPh_3, $Ph_2PCH_2CH_2PPh_2$ (dppe), PMe_2Ph, $PMPh_2$ and $P(OMe)_3$] with [60]fullerene-Pd (or Pt) complexes (or polymers).[31] The NMR spectra of some of the derivatives give evidence of fluxional behaviour (which is sometimes termed a 'haptotropic rearrangement') as noted for other compounds above and in Sec. 11.3. In the dppe compound, the palladium, phosphorus, and carbon atoms of the methylene groups form a ring. A similar arrangement involving platinum results from reacting $Pt(1,5$-cyclo-ocatadiene)$_2$ with the diphosphines [$Ph_2P(CH_2)_nPPh_2$, $n = 2,3$]

to give the corresponding platinum-phosphine complexes that form a monoadduct with [60]fullerene.[32]

A more extensive range of phosphite derivatives of [60]fullerene, $C_{60}[M(P(OR)_3)_2]_n$, M = Pt, Pd, Ni, R = Ph, Bu, Et, n = 1,2, can be made by reaction of the metal phosphite complex with the fullerene.[33] Solutions of the products are green, with no indication of formation of orange-red solutions that are characteristic of the hexa-adducts obtained with the phosphine-derived ligands. The nickel derivatives are very air sensitive in both solid and solution, whereas the Pd and especially the Pt derivatives in solution decompose only slowly over periods of weeks.

11.5 Fullerene Complexes of Iron and Ruthenium

Reaction of [60]fullerene with $Fe_2(CO)_9$ produces $(\eta^2\text{-}C_{60})Fe(CO)_4$ in high yield.[22] Formation of one ruthenium derivative has been described above (Sec. 11.3). A very interesting (and first) hexahapto fullerene complex is obtained by the reaction of [60]fullerene with $Ru_3(CO)_{12}$.[34] The corresponding reaction with [70]fullerene takes place at the hexagon adjacent to the polar pentagon, as expected, giving $(\mu_3\text{-}\eta^2,\eta^2,\eta^2,\text{-}C_{70})Ru_3(CO)_9$, resulting from addition across the 1,2-, 3,4- and 5,6-bonds, **11.13**. Moreover, di-addition produces three isomers (1:2:3 ratio), which is very close to that expected (1:2:2 ratio) from addition to either of the hexagons surrounding the other polar pentagon. The most abundant isomer is **11.14**, $(\mu_3\text{-}\eta^2,\eta^2,\eta^2,\text{-}C_{70})_2 Ru_3(CO)_9$, resulting from addition across the 1,2-, 3,4-, 5,6-, 41,58, 42,43- and 59,60-bonds.

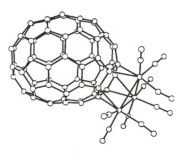

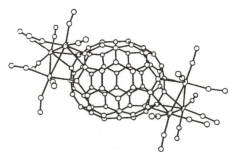

11.13   **11.14**

Another indication of the fundamental alkene-like property of fullerenes comes from reaction of the ruthenium complex, $\{[(\eta^5\text{-}C_5Me_5)Ru(CH_3CN)_2]_3^+\}$ $(O_3SCF_3^-)$ with [60]fullerene which results in displacement of only one CH_3CN group by the fullerene, a reaction characteristic of alkenes, whereas by contrast, aromatics readily displace all three such ligands.[27,35]

11.6 Fullerene Complexes of Molybdenum, Vanadium and Tantalum

[60]Fullerene forms metallocene complexes of the general formula $(\eta^2\text{-}C_{60})[M\eta\text{-}(C_5H_5)_2]$, $M = \text{Mo}^{22}$ or V^{36} by displacement of hydrogen from the precursor in the first case, and by direct reaction with vanadocene in the latter; a molybdenum derivative with n-butyl groups attached to the cyclopentadienyl rings is also known.[22] Each complex is rather unstable, the molybdenum compound decomposing rapidly in THF, whilst the vanadium complex (obtained at 240 K) dissociates back to vanadocene and [60]fullerene at room temperature, i.e. the formation is a reversible process.

A related cyclopentadienyl complex, $(\eta^2\text{-}C_{60})[TaH(\eta\text{-}C_5H_5)_2]$, can be made by displacement of H_2 from the corresponding trihydro precursor.[22]

11.7 Miscellaneous Fullerene-Metal Complexes

Reaction of [60]fullerene with $Re_2(CO)_{10}$ in benzene produces $C_{60}[Re(CO)_5)]_2$ which is the only compound obtained so far that has σ-bonding between the metal and the cage. The compound is unstable and dissociates back to starting materials with a first-order rate coefficient of 7.4×10^{-5} s^{-1} at room temperature. The structure is believed to be as shown in **11.15**, and it may also be produced by displacement of Ph_3C from $(\eta^3\text{-}Ph_3C)Re(CO)_4$ in the presence of CO.[37]

A unique cyclopentadienyl complex exists in which the cyclopentadiene moiety comes from [60]fullerene itself. Thus reaction of TlOEt in THF with $C_{60}Ph_5H$ produces dark red plates of the thallium(I) complex, $Tl(\eta^5\text{-}C_{60}Ph_5)\cdot2.5THF$, **11.16** (location of THF omitted).[38] Likewise, treatment of $C_{60}Ph_5H$ with either LiOtBu, KOtBu or $Cu(OtBu)(PEt_3)$ gives corresponding derivatives with Li, K, and Cu coordinated to the cyclopentadienyl ring. Many other compounds of this type (involving other aromatic rings surrounding the pentagonal ring may ultimately be synthesised.

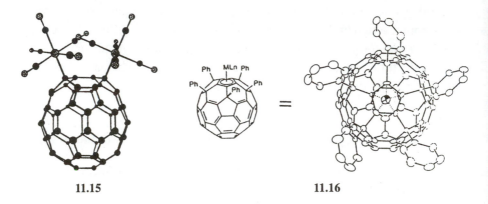

11.15   **11.16**

The ions $C_{60}Fe^+$ and $C_{70}Fe^+$, in which the iron is externally bound, are produced (within a mass spectrometer) by ligand exchange between the fullerene and $Fe(C_nH_{2n})^+$ ($n = 2-5$), the latter obtained by reaction between laser desorbed Fe^+ and pentane; Co, Ni, Cu, Rh, and La have also been similarly externally bound to fullerenes.[39] These compounds are capable of undergoing ligand exchange reactions, e.g. CoC_{60}^+ reacts in the gas phase with propane to give, after collision-induced dissociation, $C_{60}(CH_2)_{1,2}^+$.[40] This utilises ion cyclotron mass spectrometry, in which individual species ions can be studied by being selectively held within an electric field.

Reaction of $Fe(benzyne)^+$ with [60]fullerene gives $FeC_{60}C_6H_4^+$, which is believed to have structure **11.17**. $Fe(biphenylene)^+$ reacts similarly, and these compounds undergo collision-induced dissociation as above to give e.g. $C_{60}(C_6H_4)^+$.[41]

Under UV irradiation, $Fe_2S_2(CO)_6$ reacts with [60]fullerene to give the complexes $C_{60}[S_2Fe_2(CO)_6]_n$, $n = 1-6$, compounds with $n = 1-3$ having been isolated and characterised; **11.18** shows the structure for $n = 1$. The same reaction with [70]fullerene produces the complexes $C_{70}[S_2Fe_2(CO)_6]_n$, with $n = 1-4$.[42]

It could be anticipated that strong interactions, if not actually leading to bonding, could exist between strong electron donors and fullerenes with their strong electron-acceptor properties. Ferrocene is such a donor, and mixing solutions of [60]fullerene and ferrocene in benzene results in co-crystallisation, to give $C_{60} \cdot [(\eta^5 - C_5H_5)_2Fe]_2$, the crystal packing being shown in **11.19**. One of the (electron-rich) cyclopentadienyl rings of ferrocene is in slightly offset

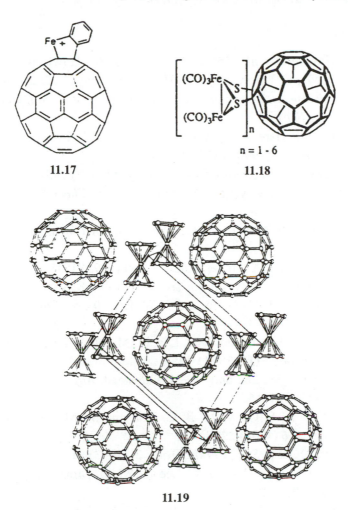

11.17

$n = 1 - 6$

11.18

11.19

π-contact with one of the (electron-poor) pentagonal rings of the fullerene. The inter-plane distance (3.3 Å), similar to that between the planes of graphite, is typical of that between neutral species, rather than between ions, so that full electron transfer does not occur.[43] A similar π,π-contact between cyclopentadienyl rings is found in the ternary material $C_{60}\cdot[Fe_4(CO)_4(\eta^5\text{-}C_5H_5)_4]\cdot3C_6H_6$ obtained by cocrystallisation between [60]fullerene and $Fe_4(CO)_4(\eta^5\text{-}C_5H_5)_4$ in benzene. Here three of the four rings in the complex interact with an adjacent fullerene cage.[44]

References

1. J. M. Hawkins *et al.*, **55** (1990) 6250; J. M. Hawkins, A. Meyer, T. A. Lewis, S. Loren and F. J. Hollander, *Science*, **252** (1991) 312; J. M. Hawkins, S. Loren, A. Meyer and R. Nunlist, *J. Am. Chem. Soc.*, **113** (1991) 7770.
2. R. Taylor, J. P. Hare, A. K. Abdul-Sada and H. W. Kroto, *J. Chem. Soc.*, *Chem. Commun.*, (1990) 1423.
3. J. M. Hawkins *et al.*, *J. Am. Chem. Soc.*, **114** (1992) 7954.
4. J. M. Hawkins, A. Meyer and M. Nambu, *J. Am. Chem. Soc.*, **115** (1993) 9844.
5. J. M. Hawkins, A. Meyer and M. A. Solow, *J. Am. Chem. Soc.*, **115** (1993) 7499.
6. J. M. Hawkins and A. Meyer, *Science*, **260** (1993) 1918.
7. R. Taylor, *J. Chem. Soc.*, *Perkin Trans. 2*, (1993) 813.
8. A. L. Balch, V. J. Catalano and J. W. Lee, *Inorg. Chem.*, **30** (1991) 3980.
9. A. L. Balch, V. J. Catalano, J. W. Lee, M. M. Olmstead and S. R. Parkin, *J. Am. Chem. Soc.*, **113** (1991) 8953.
10. A. L. Balch, J. W. Lee and M. M. Olmstead, *Angew. Chem. Intl. Edn Engl.*, **31** (1992) 1356.
11. A. L. Balch, A. S. Ginwalla, B. C. Noll and M. M. Olmstead, *J. Am. Chem. Soc.*, **116** (1994) 2227.
12. R. Taylor, *J. Chem. Soc.*, *Perkin Trans. 2*, (1993) 813.
13. A. L. Balch, V. J. Catalano, J. W. Lee and M. M. Olmstead, *J. Am. Chem. Soc.*, **114** (1992) 5455.
14. A. L. Balch, J. W. Lee, B. C. Noll and M. M. Olmstead, *J. Am. Chem. Soc.*, **114** (1992) 10984; *Inorg. Chem.*, **33** (1994) 5238.
15. M. Rasinkanfas, T. T. Pakkanen, T. A. Pakkanen, M. Ahlgren and J. Rouvinen, *J. Am. Chem. Soc.*, **115** (1993) 4901.
16. A. V. Usatov, A. L. Blumenfeld, E. V. Voronstov, L. E. Vonogradova and Yu. N. Novikov, *Mendeleev. Commun.*, (1993) 229; A. V. Usatov, E. V. Voronstov, L. E. Vonogradova and Yu. N. Novikov, *Russ. Chem. Bull.*, **43** (1994) 1572.
17. S. Schreiner, T. Gallaher and H. K. Parsons, *Inorg. Chem.*, **33** (1994) 3021.

18. R. S. Koefod, M. F. Hugdens, and J. R. Shapley, *J. Am. Chem. Soc.*, **113** (1991) 8957.
19. A. L. Balch, D. A. Costa, J. W. Lee, B. C. Noll and M. M. Olmstead, *Inorg. Chem.*, **33** (1994) 2071.
20. A. L. Balch, D. A. Costa, B. C. Noll and M. M. Olmstead, *J. Am. Chem. Soc.*, **117** (1995) 8926.
21. A. L. Balch, J. W. Lee, B. C. Noll and M. M. Olmstead, *Inorg. Chem.*, **32** (1993) 3577.
22. R. E. Douthwaite, M. L. H. Green, A. H. H. Stephens and J. F. C. Turner, *J. Chem. Soc., Chem. Commun.*, (1993) 1522.
23. Y. Ishi, H. Hoshi, Y. Hamada and M. Hidai, *Chem. Lett.*, (1994) 801.
24. M. L. H. Green and A. H. H. Stephens, *Chem. Commun.*, (1997) 793.
25. P. J. Fagan, J. C. Calabrese and B. Malone, *Science*, **252** (1991) 2252.
26. V. V. Bashilov, P. V. Petrovskii, V. I. Sokolov, S. V. Lindeman, I. A. Guzey and Y. T. Struchkov, *Organometallics*, **12** (1993) 991.
27. P. J. Fagan, J. C. Calabrese and B. Malone, *Accounts Chem. Res.*, **25** (1992) 134.
28. P. J. Fagan, J. C. Calabrese and B. Malone, *J. Am. Chem. Soc.*, **113** (1991) 9408.
29. R. Taylor, *J. Chem. Soc., Perkin Trans. 2*, (1992) 1667.
30. A. L. Balch, L. Hao and M. M. Olmstead, *Angew. Chem. Intl. Edn. Engl.*, **35** (1996) 188.
31. V. V. Bashilov, P. V. Petrovskii and V. I. Sokolov, *Russ. Chem. Bull.*, (1993) 392; H. Nagashima, H. Yamaguchi, Y. Kato, Y. Saito, M. Haga and K. Itoh, *Chem. Lett.*, (1993) 2153; H. Nagashima *et al.*, *Chem. Lett.*, (1994) 1207.
32. M. van Wijnkoop *et al.*, *J. Chem. Soc., Dalton Trans.*, (1997) 675.
33. F. J. Brady, D. J. Cardin and M. Domin, *J. Organometal. Chem.*, **491** (1995) 169; S. A. Lerke, B. A. Parkinson, D. H. Evans and P. J. Fagan, *J. Am. Chem. Soc.*, **114** (1992) 7807; H. Nagashima, H. Yamaguchi, Y. Kato, Y. Saito, M. A. Haga, K. J. Itoh, *Chem. Lett.*, **12** (1993) 2153.
34. H. Hsu and J. R. Shapley, *J. Am. Chem. Soc.*, **118** (1996) 9192; H. Hsu, S. R. Wilson and J. R. Shapley, *Chem. Commun.*, (1997) 1125.
35. P. J. Fagan, J. C. Calabrese and B. Malone, *Science*, **252** (1991) 1160.
36. V. K. Cherkasov, Yu. F. Rad'kov, M. A. Lopatin and M. N. Bochkarev, *Russ. Chem. Bull.*, **43** (1994) 1834.

37. S. Zhang, T. L. Brown, Y. Du and J. R. Shapley, *J. Am. Chem. Soc.*, **115** (1993) 6705.

38. M. Sawamura, H. Iikura and E. Nakamura, *J. Am. Chem. Soc.*, **118** (1996) 12850.

39. L. M. Roth *et al.*, *J. Am. Chem. Soc.*, **113** (1991) 6298; Y. Huang and B. S. Frieser, *J. Am. Chem. Soc.*, **113** (1991) 9418.

40. S. Z. Kan, Y. G. Byun and B. S. Frieser, *J. Am. Chem. Soc.*, **116** (1994) 8815.

41. S. Z. Kan, Y. G. Byun and B. S. Frieser, *J. Am. Chem. Soc.*, **117** (1995) 1177; S. Z. Kan, Y. G. Byun, S. A. Lee and B. S. Frieser, *J. Mass Spectrom.*, **30** (1995) 194.

42. M. D. Westmeyer, C. B. Galloway and T. B. Rauchfuss, *Inorg. Chem.*, **33** (1994) 4615; M. D. Westmeyer, T. B. Rauchfuss and A. K. Verma, *Inorg. Chem.*, **35** (1996) 7140.

43. J. D. Crane, P. B. Hitchcock, H. W. Kroto, R. Taylor and D. R. M. Walton, *J. Chem. Soc., Chem. Commun.*, (1992) 1764.

44. J. D. Crane and P. B. Hitchcock, *J. Chem. Soc., Dalton Trans.*, (1993) 2537.

12

Polymers, Dendrimers, Dimers, Dumb-bells and Related Structures

The utilisation of fullerenes in polymers has been a major goal of fullerene chemists and materials scientists. Three possibilities arise:

1. The polymer may consist of linked fullerenes (linked either directly or by means of a small 'spacer'), giving a 'pearl necklace' structure.

2. The fullerenes may be attached regularly to a polymeric backbone chain, giving a 'pendant chain' or 'charm-bracelet' effect.

3. The polymer may have the fullerene as a relatively small percentage of the overall composition, the fullerene being located at structurally significant points. The latter group includes the copolymers.

Until fullerenes can be produced at a cost of as few pence per kg, only the latter class of polymers have any prospect of being utilised at present. Good progress has been made in creating polymers showing promise of having special applications.

12.1 'Pearl-Necklace' Polymers

12.1.1 *Directly-Linked Fullerenes*

These polymers are difficult to characterise properly, and very difficult to process. Though they provide some theoretical and mechanistic interest it seems improbable that they will provide materials of interest in the short or even long term.

Fig. 12.1 Hypothetical 'pearl-necklace' polymer involving single-bond linkages.

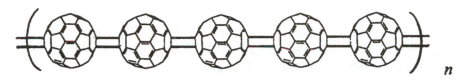

Fig. 12.2 Conjectured structure for a 'pearl-necklace' polymer involving two-bond linkages (the orthorhombic one-dimensional polymeric phase).

The simplest pattern that can be envisaged here is shown in Fig. 12.1, but is unknown, and in view of the ease of oxidation of hydrofullerenes to fullerenols (Chap. 4), may remain so. The type shown in Fig. 12.2 can be produced by irradiation of the fullerene by either UV or visible light in the absence of oxygen (which quenches the first excited triplet state of [60]fullerene and hence reaction). This results in a [2 + 2] cycloaddition of up to twelve [60]fullerene molecules.[1] The product is insoluble in toluene, but can be dissolved in boiling isodurene, which highlights the processing difficulties inherent in such materials. Given that there are thirty double bonds in [60]fullerene, numerous conformations of the product are possible, but for any given cage twelve of these are sterically precluded, and the remaining seventeen is reduced to seven through symmetry. Nevertheless, the number of conformations approaches 7^n, where n is the number of fullerene molecules involved. (Steric interactions will preclude some of the conformations that involve doubling back of the chain.) The polymer reverts to monomeric [60]fullerene on heating to *ca.* 100°C.

Polymerisation here is accompanied by contraction in the unit cell volume,[2] loss of the Raman band at 1469 cm^{-1} and formation of one at 1458 cm^{-1}, which is consistent with the altered bonding in the product. Polymerisation only occurs to any significant extent above 260 K, the temperature at which [60]fullerene becomes disordered thereby permitting close and parallel approach of the bonds required for the [2 + 2] cycloaddition process, and which calculations have indicated to be the most plausible mechanism for dimerisation and polymerisation of fullerenes.[3]

Linkage of [60]fullerene to give an orthorhombic one-dimensional polymer can also be brought about by the use of very high pressures of 5 gigapascals and temperatures between 500 and 800°C, which gives a toluene-insoluble material with a 9.22 Å intermolecular spacing;[4] pressure polymerisation causes an additional band to appear in the Raman spectrum at 1463 cm⁻¹.[5] The fullerene centre-to-centre distance of 9.22 Å was perceived to be a problem since calculations predicted a value of 8.5 Å for $(C_{60})_2$,[6] a prediction confirmed by the published data for this molecule, **12.1**, which has been synthesised recently.[7] However, in the polymer, the cages will be elongated along the direction of the polymer chain, due to the presence of sp^3 carbons on opposite sides of each cage, and this should give rise to a longer overall centre-to-centre distance, as observed. Such features may also account for the signals for the sp^3-hybridised carbons appearing at 76.2 ppm for $(C_{60})_2$ compared to 73.5 ppm for the polymer.[8]

Polymerisation under pressure also produces two other phases,[9] one being tetragonal, the other (major) one being rhombohedral two-dimensional, **12.2**, and having a lattice parameter of 9.19 Å, similar to that for the one-dimensional phase. Although the structure as drawn appears planar, it is actually crown-shaped; the material behaves as a semiconductor.[10]

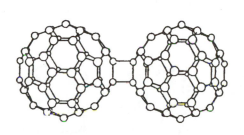

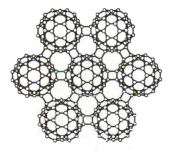

12.1   12.2

[70]Fullerene also polymerises under photochemical irradiation, but here only four molecules become detectably linked, a result attributed to there being fewer reactive bonds in [70]fullerene compared to [60]fullerene, and fewer conformations available that can result in reaction.[11] [70]Fullerene is also generally less reactive than [60]fullerene towards cycloaddition reactions. Polymerisation of [70]fullerene results in *expansion* in the unit cell volume so that pressure polymerisation is unfavourable.[2]

Further information concerning the polymers derived from [60]fullerene have come from drift-tube experiments, in which the rates at which vaporised molecular species pass through a region under electrostatic influence is measured; species of different size and shape require different times. These experiments indicate two distinct species are involved. One corresponds to structures resulting from [2 + 2] cycloaddition, the other to a more fused but unspecified isomer.[12] If higher powers for the desorbing laser are used, then species C_{118}, C_{116} etc. arising from C_2 loss are obtained, and similar losses occur from species arising from higher levels of polymerisation.[13,14] For example, the peak corresponding to $10\,C_{60}$ is 500 amu deficient, which indicates a loss of 42 carbon atoms. The energies for possible dimer structures have been calculated and show for example, that the dispirane, **12.3** could be the structure for C_{118}.

Notably, C_{119} (**12.4**) made by heating a mixture of $C_{60}O$ and C_{60},[15] has been characterised and is a spirane.[16] The proposed formation mechanism involves elimination of CO from $C_{60}O$, giving a carbene which can either add across a 6,6-bond,[17] or insert into a 6,5-bond of another C_{60} molecule. Subsequent [2 + 2] cycloaddition between the cages gives rise to two products,[18] but that deriving from the 6,5 insertion is sterically less hindered than that derived from 6,6-addition, so is the preferred product.[16,19] (see also Sec. 9.1.1). C_{129}

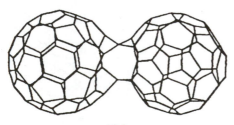

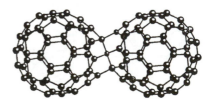

12.3   12.4

(derived from $C_{70}O/C_{60}$, or the converse) and C_{139} (derived from $C_{70}O/C_{70}$) have also been detected by mass spectrometry,[15] but not isolated yet. C_{119} can also be made by heating $C_{60}O$ alone and in this case CO_2 is eliminated, thus indicating an additional formation mechanism.[20]

The rate of photodimerisation of [60]fullerene increases by three orders of magnitude if the reaction is carried out under an atmosphere of NO, though extended irradiation causes degradation. The NO is believed to react with the diradical formed in the primary photochemical step, thereby producing a more stable (mono) radical which then attacks a second [60]fullerene molecule.[21] Photoirradiation of acetonitrile-toluene solutions of fullerene clusters (obtained by addition of acetonitrile to toluene solutions of fullerenes) causes them to polymerise, but characterisation of the polymers is difficult;[22] the average size of 156 nm suggests that at least 100 [60]fullerene cages are involved.

12.1.2 Indirectly-Linked Fullerenes

The next most simple polymeric structure that can be envisaged is one involving a spacer group, e.g. Fig. 12.3, which shows a one-dimensional linkage. Linkage may also be in two- or three dimensions, termed net- and lattice polymers by the writer.[23] Unless the addition on the cage occurs across either the 1,2- or 1,4-positions, hydrogen atoms (or some other group) will be necessary in order to avoid an exceptional amount of destabilising bond re-organisation in the cages (with numerous double bonds in pentagons in each cage).

Localised addition must be involved in the incorporation of fullerenes into a forming polymer chain, a procedure which gives rather low control of the polymer product. For example, pyrolysis of paracyclophane produces the *para*-xylylene diradical which can then be polymerised in the presence of [60]fullerene to give the polymer of formula $-\{([60]\text{fullerene})_n\text{-}(CH_2C_6H_4$

Fig. 12.3 Hypothetical 'pearl-necklace' polymer, possessing a spacer group.

$CH_2)_m\}_x$ — in which the m/n ratio is *ca*. 3.4. Due to extensive cross linking, the product is insoluble, and it is also unstable in air.[24]

Localised addition is likely in the fullerenation of polycarbonate. The latter, under UV irradiation, is assumed to cleave across the ~ OCO-O ~ bond giving radicals ~ $OCO^\bullet$ and $^\bullet O$ ~ which then attach to the cage. In this simple method, up to 6.3 wt.-% of [60]fullerene is incorporated into the resultant polymer.[25]

Both [60]- and [70]fullerenes can be incorporated into polystyrene through free-radical copolymerisation of styrene and methyl methacrylate in the presence of the fullerene,[26] the possible reaction steps being as shown e.g. in Eqs. 1 and 2 (PS = polystyrene); localised addition must be involved here too. The intermediate $PSC_{60}^\bullet$ can in principle propagate by reacting with another styrene monomer, (Eq. 3) or react with another $PS^\bullet$ to terminate the reaction (Eq. 4) and in this latter case there will be only one fullerene molecule per polymer chain. This latter is largely the case, though the number of fullerenes per chain appears to increase with molecular weight of the chain, indicating that the propagation step, plays a role. Fullerene incorporations are in the range 0.21–5.1% but yields (10–85%) vary according to the fullerene incorporation, being lower when this is higher, which suggests that the radical-trap behaviour of the fullerenes is responsible. The inhibiting effect of [60]fullerene on radical polymerisation of these and other alkenes has been confirmed, though in the case of polymerisation of both cyanovinyl acetate and 2-cyanoethyl acetate, good yields are obtained for reasons as yet unclear. An electronic stabilisation of the intermediate radical by the electron-withdrawing CN group may be responsible.[27]

$$\text{AIBN} + \text{styrene} \quad \rightarrow \quad \text{PS}^\bullet \tag{1}$$

$$\text{PS}^\bullet + C_{60} \quad \rightarrow \quad \text{PS-}C_{60}^\bullet \tag{2}$$

$$\text{PS-}C_{60}^\bullet + \text{PS} \quad \rightarrow \quad \text{PS-}C_{60}\text{-PS}^\bullet \tag{3}$$

$$\text{PS-}C_{60}^\bullet + \text{PS}^\bullet \quad \rightarrow \quad \text{PS-}C_{60}\text{-PS} \tag{4}$$

The polymer $C_{60}Pd_n$, is formed from [60]fullerene and the complex Pd_2(dibenzylideneacetone)$_3$CHCl$_3$ (which undergoes ligand displacement).[28] With [60]fullerene in excess, n is never less than one, consistent with a chain having one palladium per fullerene cage. On heating this material, some [60]fullerene is released, and the value of n increases to *ca*. 3, indicating the

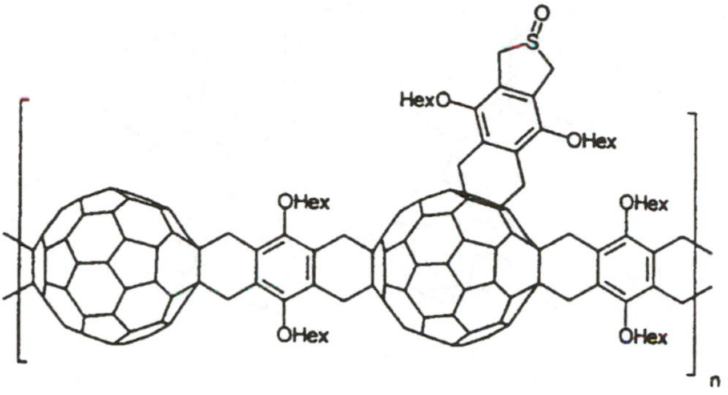

12.5

formation of a lattice polymer. For n values $> ca.$ 3, the polymer behaves as a catalyst for hydrogenation, suggesting the presence of some palladium on the polymer surface. Similar polymers are formed with platinum.[29]

Localised addition is not involved in formation of a [60]fullerene copolyamide because here $HCCO_2H$ [1 + 2] cycloaddends are located by a synthetic procedure across the 1,2- and 18,36-positions. Subsequent reaction with isophthalic acid and 4,4'-diaminodiphenyl ether then produces polymers with M_w values in the 50,000–60,000 region.[30]

Good progress towards solving the solubility problem has been achieved with two strategies: reacting with a bis-*o*-quinodimethane (derived from a bis-sulphone precursor — see also Sec. 9.4.1) possessing soublising hexyloxy groups, and at the same time having a mono-*o*-quinodimethane present which inhibits cross linking. With [60]fullerene this leads to a product comprised of multiples of units such as that conjectured in **12.5**. These polymers have up to 80 [60]fullerene cages incorporated and they can be processed to amorphous thin films by spin coating.[31]

12.2 'Pendant-Chain' Polymers

Embryonic 'pendant-chain' polymers have been made, **12.6** (the central aryl ring has either a *meta* or *para* linkage),[32] but with these problems of solubility arise also. Attempts have been made to overcome this by having chains of atoms connecting the aryl groups, achieved by making derivatives with hydroxy

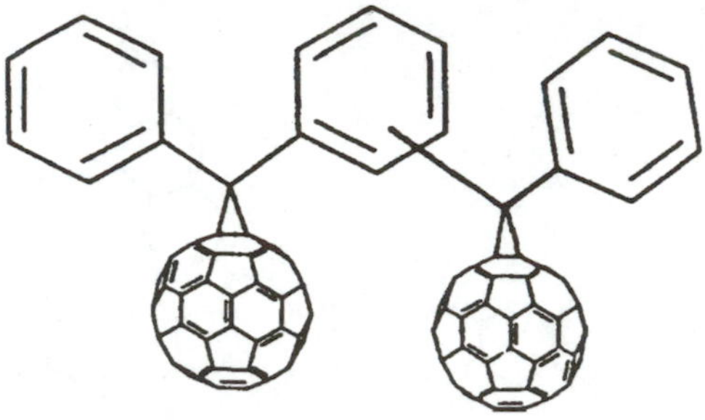

12.6

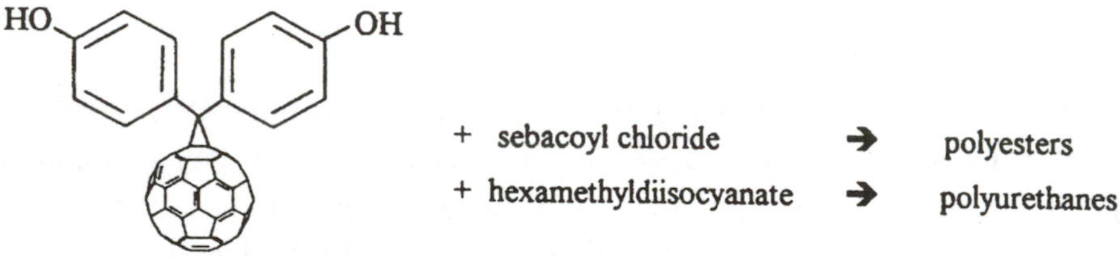

+ sebacoyl chloride ➔	polyesters
+ hexamethyldiisocyanate ➔	polyurethanes

Fig. 12.4 Routes to the formation of polyesters and polyurethanes.

groups in the aryl rings, *para* to the bridging group (Fig. 12.4). Polymerising these fullerene derivatives with sebacoyl chloride [$ClCO(CH_2)_8COCl$] gives a (more tractable) polyester, though polymerising in the presence of hexamethyldiisocyanate (to give a polyurethane) does not.[33]

In some pendant-chain polymers, the fullerene is attached to a polymer chain but the inter-fullerene distance can be considerable. The fullerene can be incorporated either while the other component of the polymer is being formed from monomer, or attached subsequently to a preformed polymer chain. Both techniques have been used and the products are more soluble and processable than those described above.

12.2.1 *Attachment of a Fullerene to a Preformed Polymer Chain*

An example is the reaction of non cross-linked polystyrene with fullerene in the presence of aluminium chloride, giving a highly cross-linked product as a result of Lewis acid-catalysed alkylation of the fullerene.[34]

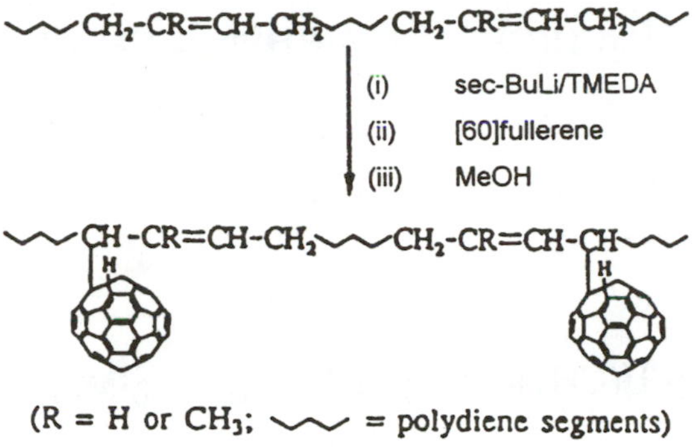

Fig. 12.5 Formation of a 'pendant-chain' polymer derived from polyalkenes.

An alternative procedure makes use of the high reactivity of fullerenes towards anions and nucleophiles. Polyethylene films, functionalised with diphenylmethyl groups, have acidic hydrogens present due to these groups. Deprotonation may therefore be effected with e.g. BuLi-tetramethylenediamine to give an anionic polythene surface which then attacks [60]fullerene giving the fullerene-incorporated polymer.[35] A related example utilises the lithiated derivatives of either *cis*-1,4-polyisoprene, or *cis*-1,4-butadiene (Fig. 12.5), giving materials with 50% and 10% w/w incorporation, respectively.[36]

The addition of up to ten polymer and ten methyl groups to a single [60]fullerene can be achieved by reactions of anions derived from polystyrene chains (Fig. 12.6).[37] Here, the [60]fullerene is inserted into a number a polymer chains, albeit at the ends of each, giving rise to a lattice or star polymer, which could also be classed as a pearl-necklace type. There is a trend to the formation of species of higher molecular mass as the polystyrene:[60]fullerene ratio increases, and appears to pass through a maximum for values of $x = 4-10$, which infers that this is the maximum number of chains that can be accommodated on the fullerene core, (which is consistent with conclusions derived from other work). The product is highly soluble and melt processable, making it amenable to spin coating, solvent casting, and melt extrusion.

The number of polystyrene chains attached in this way to a given fullerene can be limited to six through end-capping the chain with additional phenyl groups. This provides greater delocalisation of the anionic charge, so reducing its reactivity towards the cage.[38]

$$CH_3CH_2CH(CH_3)\text{-}(\text{-}CH_2CHPh\text{-})_n\text{-}CH_2CHPh^-Li^+$$

$$\downarrow \text{ aq. THF}$$

$$\downarrow C_{60}$$

$$\downarrow \text{MeI}$$

$$[CH_3CH_2CH(CH_3)\text{-}(\text{-}CH_2CHPh\text{-})_n]_x\text{-}C_{60}\text{-}Me_x \quad (x = 1 - 10)$$

Fig. 12.6 Formation of a [60]fullerene-polystyrene copolymer.

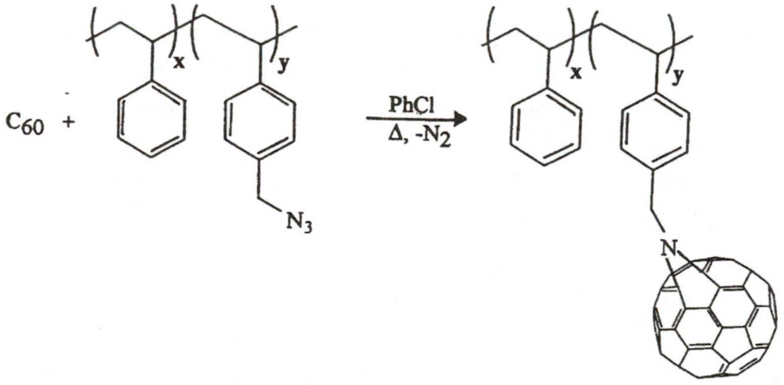

Fig. 12.7 Soluble pendant-chain polymers formed from [60]fullerene and azido-substituted polystyrenes.

Various strategies can be adopted in order to improve the solubility of these polymer types, usually by the incorporation of nitrogen, or amino groups. For example, [60]fullerene cycloadds to azido-substituted polystyrenes to give the polymers shown in Fig. 12.7. In these, x is either 99, 94, or 89, with corresponding y values of 1, 6 and 11, the respective [60]fullerene incorporations being 5.5, 21 and 29 wt.-%. The molecular weights range from 27,000 (polydispersity 2.04) to 38,500 (polydispersity 3.12), the respective glass transition temperatures being 112, 142 and 160°C, compared to 97°C for azidopolystyrene, and the polymers have high solubility in organic solvents.[39]

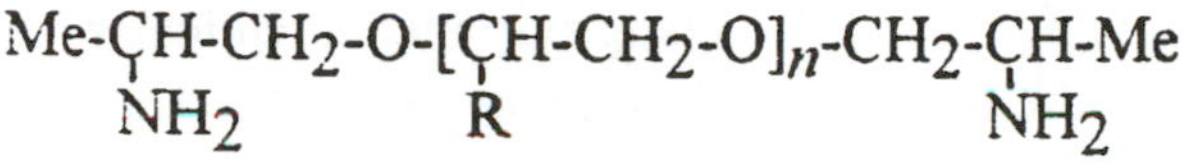

$$Me\text{-}\underset{NH_2}{CH}\text{-}CH_2\text{-}O\text{-}[\underset{R}{CH}\text{-}CH_2\text{-}O]_n\text{-}CH_2\text{-}\underset{NH_2}{CH}\text{-}Me$$

Fig. 12.8 Aminopoly(oxyalkene) polymers (R = H, Me, *n* variable) which nucleophilically add to [60]fullerene to give pendant-chain derivatives.

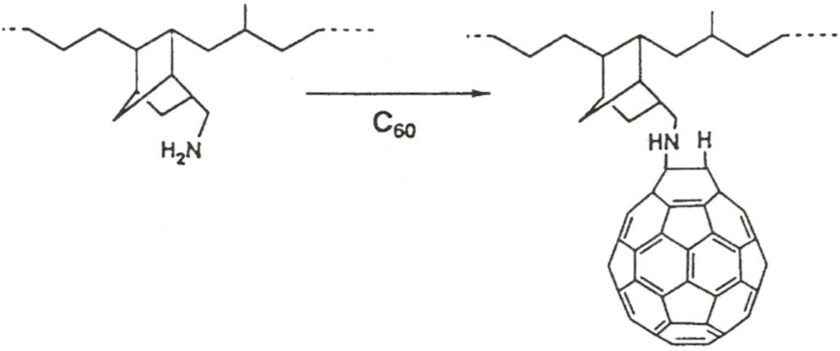

Fig. 12.9 Reaction between [60]fullerene and amine-functionalised ethylene-propylene copolymer.

[60]Fullerene reacts with amino-containing poly(oxyethylene) or poly(oxypropylene) polymers (Fig. 12.8, R = H, Me, *n* variable) to give toluene- and water-soluble products (for the latter, acid addition is necessary in some cases), and an average of *ca.* three polymer molecules are joined to each fullerene cage (so that the product is dendritic). When the aminopolymer:[60]fullerene ratio is small, the opportunity for cross-linking the fullerene is less, so soluble products are obtained (albeit slowly). At higher ratios, soluble products are obtained only during the first 30 h of reaction, after which cross-linking evidently occurs.[40]

This technique (in which a polymer with primary amino groups in the side chain goes 'fishing' for [60]fullerene) has been used to make some other polymers which have good solubility and up to 20 wt.-% fullerene incorporation.[41] Similarly, reaction (Fig. 12.9) between [60]fullerene and primary amine functionalised ethylene-propylene terpolymer gives a polymer 95% of which is soluble in xylene, the insoluble residue being attributed to cross-linked material.[42] Yet another variation employs an amino-terminated polystyrene in which the amino end group nucleophilically adds to the cage, and this lead to the formation of fullerene-end-capped polystyrenes having molecular weights ranging from 2,040 to 42,900, the corresponding glass transition temperatures being 65 and 91°C.[43]

12.2.2 *Polymers Formed from Addended Fullerenes*

Some of the most promising polymers have been made by reaction of [60]fullerene cages which have either a definite or restricted number of addends attached, and this results in greater control over the products and a narrower molecular weight range. This technique is likely to gain in importance as methods of controlled functionalisation of the cages improves.

An example is the reaction of poly(propionylethylenimine-*co*-ethylenimine) copolymer with methano[60]fullerene dicarboxylic acid to give an amide-linked polymer (Fig. 12.10) having a 12% [60]fullerene content, average molecular weight of 49,000 (polydispersity 2.67), and a high (> 90 mg ml^{-1}) aqueous solubility.[44]

In another approach, which yields a lattice polymer (see Sec. 12.1.2), the amino group, in the form of 2-methylaziridine, is attached (approximately 10-fold) to the [60]fullerene cage. This then reacts with phenolic compounds such as Novolac and Bisphenol A, or epoxides such as Epon 828, to give highly cross-linked polymers which have good friction and wear properties, as well as thermal stability.[45]

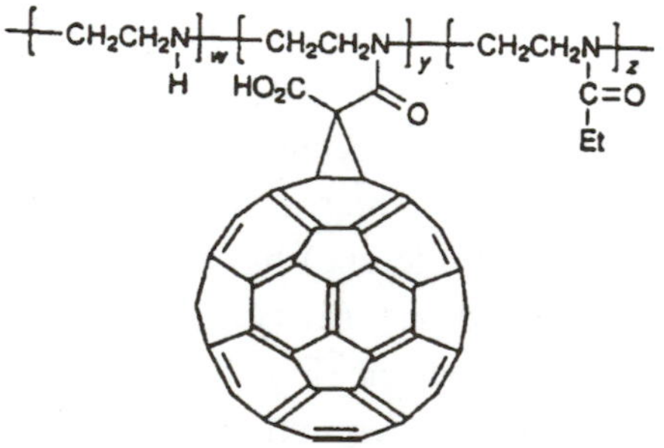

Fig. 12.10 A highly water-soluble pendant [60]fullerene poly(propionylethylenimine-*co*-ethylenimine) polymer.

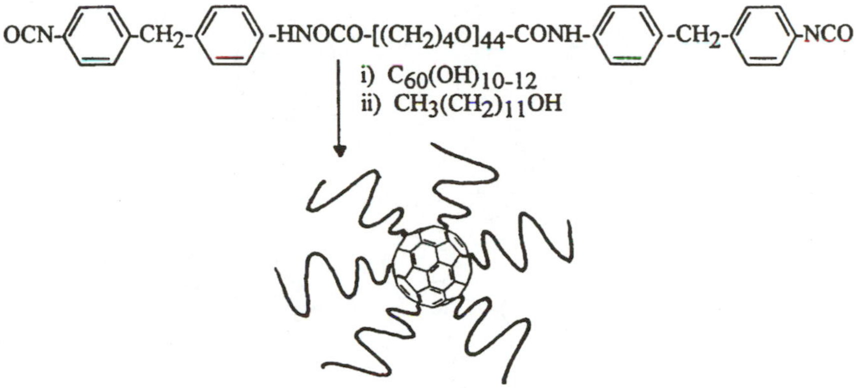

$$OCN-\langle\!\!\bigcirc\!\!\rangle-CH_2-\langle\!\!\bigcirc\!\!\rangle-HNOCO-[(CH_2)_4O]_{44}-CONH-\langle\!\!\bigcirc\!\!\rangle-CH_2-\langle\!\!\bigcirc\!\!\rangle-NCO$$

i) $C_{60}(OH)_{10-12}$
ii) $CH_3(CH_2)_{11}OH$

Fig. 12.11 Formation of a lattice polymer by reaction of [60]fullerenols with urethane-polyether prepolymer.

Some of the most successful polymers made so far are those derived from fullerenols, (which can be easily made with up 12 OH groups attached to the cage). Condensation of these fullerenols with di-isocyanated urethane polyether prepolymer, followed by quenching with dodecan-1-ol (Fig. 12.11), gives a product with an average molecular weight of 18,000 (maximum 26,100) so the polydispersity, 1.45, is extremely narrow; the glass transition temperature is −67°C. The molecular weight shows that there are six arms to each fullerene, (which are most probably located at the octahedral sites). Some hydroxy groups must remain unattacked, presumably due to steric hindrance. Notably, these polymers have greatly enhanced tensile strength, elongation, and both thermal and mechanical stability in comparison with their linear analogues or conventional polyurethane elastomers cross-linked by trihydroxylated reagents.[46] The number of attached linear polyurethane arms per [60]fullerene decreases as the molecular weight of each isocyanated polymer arm is increased.[47]

12.2.3 Star Polymers

In the above description, a number of multiple two- and three-dimensional additions to the [60]fullerene cage give net- and lattice-type polymers. These

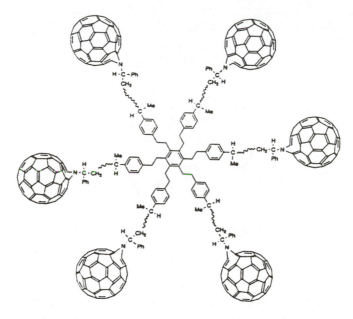

Fig. 12.12 Structure of an [60]fullerene end-capped polystyrene star polymer.

products may also be described as star polymers, in which the fullerene is the star centre. However, it is also possible to have the *polymer* as the star core, with the fullerenes as the end cap of the polymer chains, as shown in Fig. 12.12, formed by reaction of an azido-terminated star with [60][fullerene. The average molecular weight indicates that there are 30 styrene units per branch.[48]

12.3 Electroactive Polymers

Cyclic voltammetry of 61,61-bis(trimethylsilylbutadiynyl)-1,2-methano [60]fullerene, **12.7** shows a well-defined one-electron reduction at 1.00 V. The voltammetric peak increases with the number of scans during repeated potential cycling, due to the formation of an immobilised electroactive coating on the electrode, believed to be an air-stable polymer. The film continues to deposit even after it has covered the electrode suggesting that it is electrically conducting and could be used as an electrode material.[49] Electroactive

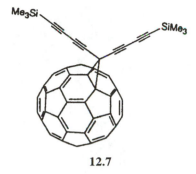

12.7

polymeric films can also be deposited on electrodes from electrochemical reduction of 1,2-epoxy[60]fullerene.[50]

12.4 Fullerenes as Polymer Dopants

The chemical incorporation of fullerenes into polymers as described above has had limited success to date. However, fullerenes can have a far more profound effect on polymers without the need to carry out chemical syntheses, and this is through addition to polymer melts. The best known property modification arising from this is the increase in photoconducting properties when fullerenes are doped into polymers containing electron-donating moieties, such as N-polyvinylcarbazole.[51] Such materials have applications as either photosensitive receptors or carrier transporting material in photocopiers.

Addition of small amounts of fullerenes (1 ~ 5 wt.-%) has a major effect on the mechanical properties of polymers. For example, [60]fullerene-doped Kevlar (poly(*p*-phenylene terephthalamide) (produced by forming the polymer in the presence of the fullerene) has a 20–50% increase in sheer modulus, and an increase in viscosity, though the tensile modulus and tenacity are reduced relative to undoped material. Addition to nylon-11 causes a *ca.* 30% increase in tenacity, though, as in the case of Kevlar, there is a reduction in the maximum elongation at the breaking load, but better resistance to breakage; the viscosity of the melt also increases. The effect of a small amount of fullerene compared to that of added carbon particles is consistent with the strong electronic interactions that are possible with the cage. A limiting factor at present is the low solubility of the fullerenes and it is likely that derivatives, which usually

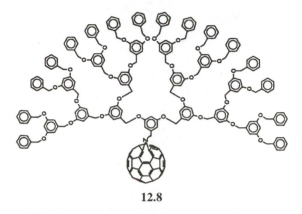

12.8

are much more soluble than the parent, will be more amenable to exploitation as polymer additives.

12.5 Dendrimers

The technique used to prepare the polymers shown in Fig. 12.4 has also been used to prepare dentritic fullerene polymers or dendrimers. In general, dendrimers have much greater solubility than polymers; for example, **12.8**, formed by cycloaddition of an azide to [60]fullerene, is very soluble in organic solvents, has a glass transition temperature of 325 K (increased relative to that of the azide precursor) and is formed with only 5% di-addition to the fullerene,[52] as might be expected in view of the bulk of the addend.

A soluble dendrimer related to the homofullerene derivative shown in Fig. 12.4, can be made from a phenolic precursor.[53] By conditions chosen to minimise side reactions (which result in increasing fullerene content with increasing molecular weight) compound **12.9**, with each phenolic oxygen of the homofullerene coupled to a dendritic unit, can be made in 90% yield.

The dendrimer **12.9** has 64 phenyl rings in the vicinity of the [60]fullerene cage. More symmetrical derivatives have also been made based on a hexakis methano[60]fullerene structure, the methano group being located in the octahedral sites since this confers aromatic stability (Sec. 3.2). In this way the fullerene core has been encased in up to 36 phenyl rings incorporated in an octahedral dendron array, **12.10**.[54] These and other dendrimers may have

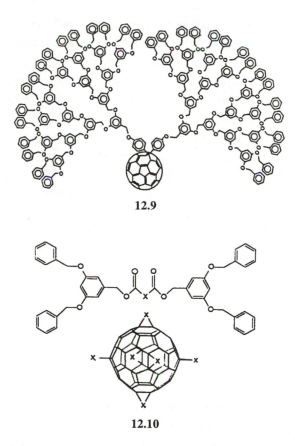

12.9

12.10

important elecro-optical properties, and amongst fullerene derivatives be the most stable since the fulleren core should be substantially protected against oxidation, a ubiquitous property of fullerenes.

12.6 Dimers

The preparation of dimers has been a goal of chemists ever since fullerenes became available, and a number have now been made.

C_{120}. Preparation of the simplest dimer, C_{120} (**12.1**) was achieved (accidentally) as a result of mechanically grinding C_{60} with KCN in a vibrating mill. The formation mechanism is believed to involve initial attack of CN^-, the resultant carbanion then attacking a second fullerene cage followed by expulsion of CN^-.[7] The inter-cage bond length (1.575 Å) is elongated relative to a normal sp^3-sp^3 bond (reflecting the mutual repulsion of the cages) and accounts for the low thermal stability of the compound, which degrades back to C_{60} on heating at 175°C for 15 min. The technique appears to be suitable for preparing various other fullerene dimers such as C_{140} etc. though of course isomers will be obtained (e.g. at least three for C_{140}). C_{120} can also be made by using lithium instead of KCN in the vibrating mill technique[55] and also by applying high pressure to $C_{60}ET_2$ [ET = bis(ethylenedithio)tetrathiafulvene].[56]

The mass spectrum (FT ion cyclotron resonance) of C_{120} shows fragmentation due to successive C_2 loss and also evidence for C_2 insertion, a feature found in some FAB mass spectra of fullerenes (see also Sec. 8.2.1.2). Preliminary investigation of the chemistry of C_{120} shows that in the Bingel reaction (Sec. 9.1.1.3) at least three derivatives are formed.

$C_{120}O$. This, the next most simple dimeric fullerene (Fig. 2.11), has been prepared by heating a mixture of C_{60} and $C_{60}O$ at 200°C for 1 h.[57] On heating to 400°C, decomposition occurs, with C_{119} (**12.4**) being one of the degradation products. The compound has a furanoid ring between the cages and thus may be named as di[60]fullereno[1,2-*b*:1,2-*d*]furan. Voltammetry studies show that the molecule undergoes six sequential one-electron reductions to to give $C_{120}O^{n-}$ where $n = 1-6$, and the addition of the first two electrons occurs successively to each fullerene cage.[58] By analogy with electron addition to [60]fullerene, it should be possible to add up to 10 electrons overall (and likewise 10 cycloaddends) under forcing conditions.

$C_{120}O_2$. The above heating procedure also produces $C_{120}O_2$, which is bis-linked by furanoid rings. The number of ^{13}C NMR peaks is consistent with two structures, but detailed analysis of the peak positions, as well as bond-order considerations showed the isolated isomer to be that given in Fig. 2.12.[59] This conclusion is also supported by theoretical calculations.[60]

12.7 Dumb-Bells

A wide variety of these structures, in which fullerenes cages are separated by a spacer, have now been synthesised:

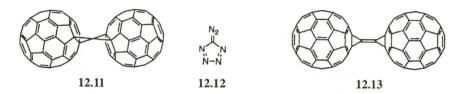

12.11   **12.12**   **12.13**

C_{121}, **12.11**, is obtained by reaction of [60]fullerene with the carbene $C_{60}C:$, produced either by removal with e.g. base of bromine from the 1,1'-dibromo-1,2-methano[60]fullerene,[61] or by decomposition of diazotetrazole **12.12** (which gives atomic carbon) in the presence of [60]fullerene.[62]

In the first of the above reactions, mass spectrometry indicates that the carbene also reacts with itself to give C_{122}, **12.13**.[61]

Structures in which the cages are separated by hexagonal rings, and produced by [4 + 2] cycloadditions, include those shown in **12.14–12.17**,[63,64] the most exotic thus far produced being **12.18**.[65] Another, produced by [3 + 2] cycloaddition is **12.19**, found to have no electronic interaction between the cages,[66] and this may be attributed to the sp³ hybridisation of the connecting carbons which therefore prevents conjugative electron relay.

Structure **12.20**, prepared by [4 + 2] cycloaddition can exist in 'extended-extended-' (**12.21**) 'extended-folded-' and 'folded-folded' (**12.22**) conformations. The structure having the latter is predicted to be to be the more stable since it brings the cages into close (2.99 Å) proximity, so that some electronic inter-cage interaction is feasible; the 'extended-extended' conformer is

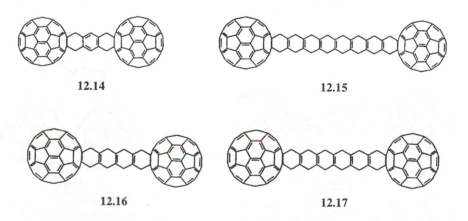

12.14   **12.15**

12.16   **12.17**

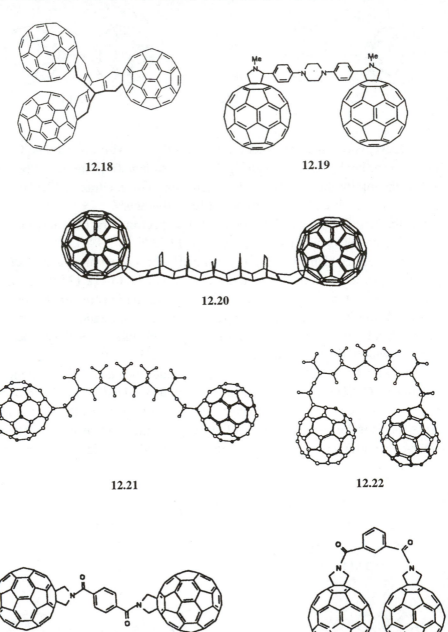

12.18

12.19

12.20

12.21

12.22

12.23

12.24

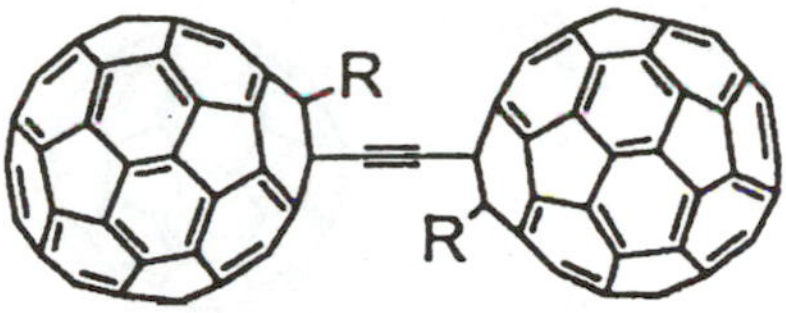

12.25 (R = $C_{12}H_{25}$) 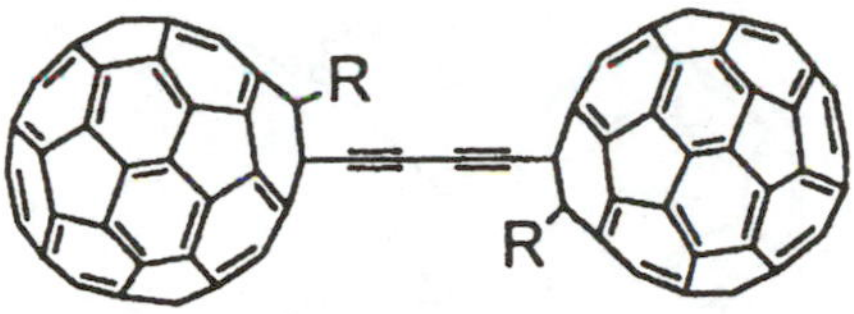   **12.26** (R = C_6H_{13}, $C_{12}H_{25}$, CH_2Ph, CH_2Othp)

particularly unstable towards mass spectrometry.[67] Two other structures with comparable conformations are **12.23** and **12.24**.[68]

Compounds with both ethyne- and a butadiyne spacers, **12.25**, **12.26**, can be made by reaction of dilithioethyne with [60]fullerene in the presence of dodecyl iodide, or by oxidatively coupling 1-alkyl-2-ethynyl-1,2-dihydro[60]fullerenes, respectively.[69,70] Electronic interaction between the cages in **12.25** is negligible, again because there are sp^3-hybridised carbons at the ends of the spacer, and the *p*-orbitals of the spacer and the cage are orthogonal. A butadiyne-spaced dumb-bell with OH groups in the 2- and 2′-positions of the [60]fullerene cages has been prepared also.[70]

A dumb-bell having a dimethanobutadiynyl spacer, **12.27**, can be produced by oxidative coupling of a mono(silyl)protected diethynylmethano[60]fullerene.[71] The most complex dumb-bells made so far are **12.28**,[72] and **12.29**[73] (a rotaxane). The former, in which the fullerene centre-to-centre distance is 16.56 Å, is prepared by reacting [60]fullerene with a *bis* iridium Vaska type compound (Sec. 11.2), that has the phosphorus ligands on each iridium linked by chains of seven methylene groups.

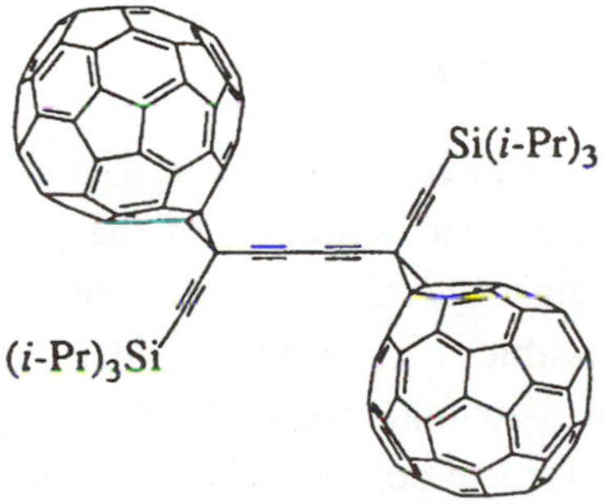

12.27

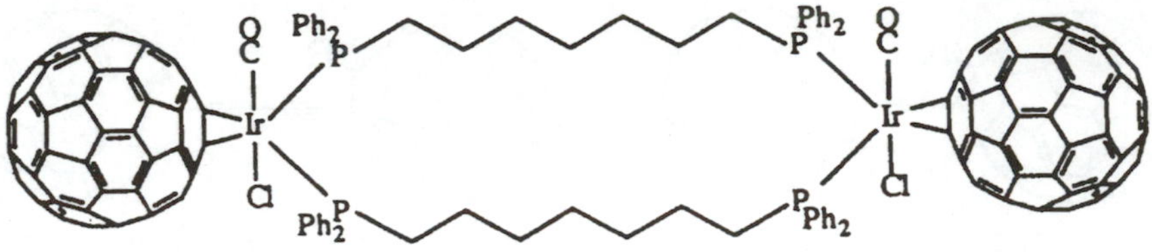

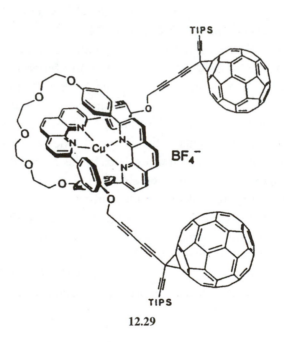

12.28

12.29

12.8 Ball and Chain Structures

Various of such structures have been prepared, generally via cycloaddition reactions, but there is no clear definition of the point at which an addend becomes a 'chain'. The objective of more recent syntheses has been investigation of the possible through-space electronic interactions between the cage and the tail of the chain, especially photoinduced electron transfer and charge separation. Typical compounds, some of which show promise in this direction are shown **12.30–12.36**.[74–76]

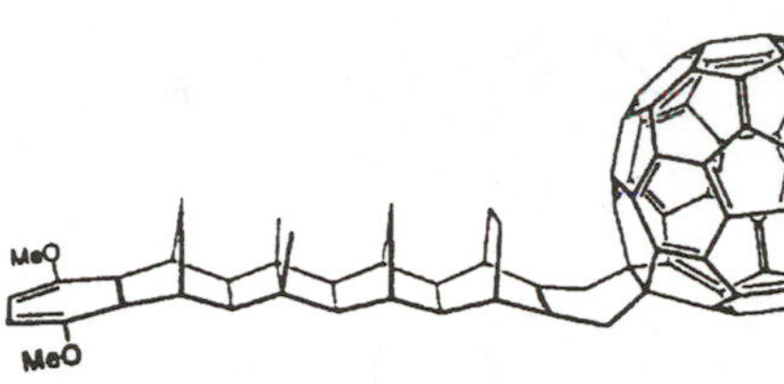

12.30

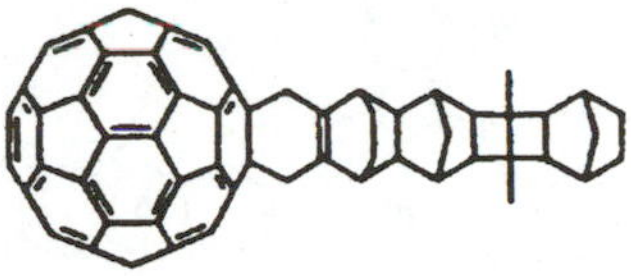

12.31

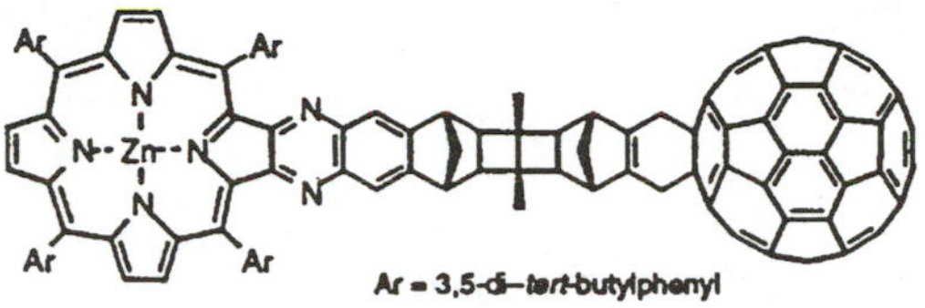

Ar = 3,5-di-*tert*-butylphenyl

12.32

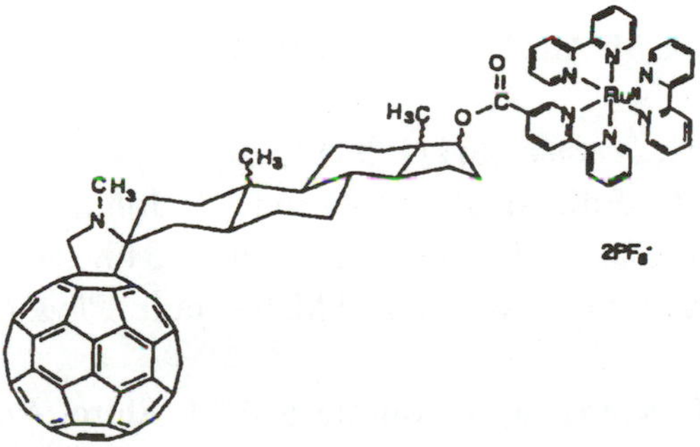

2PF$_6^-$

12.33

n = 0, 1

X = H, NMe$_2$

[PF$_6$]$_2$

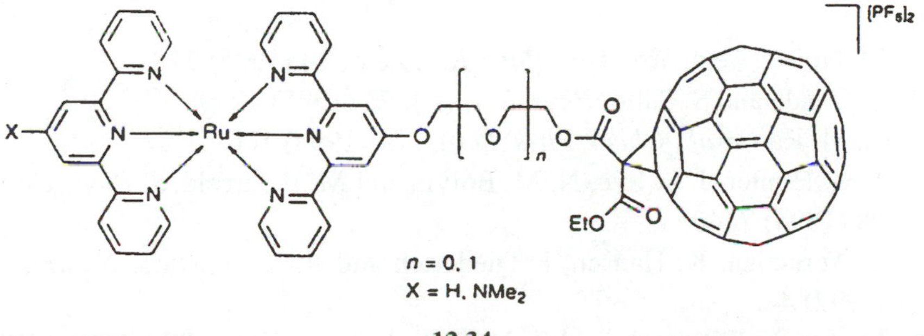

12.34

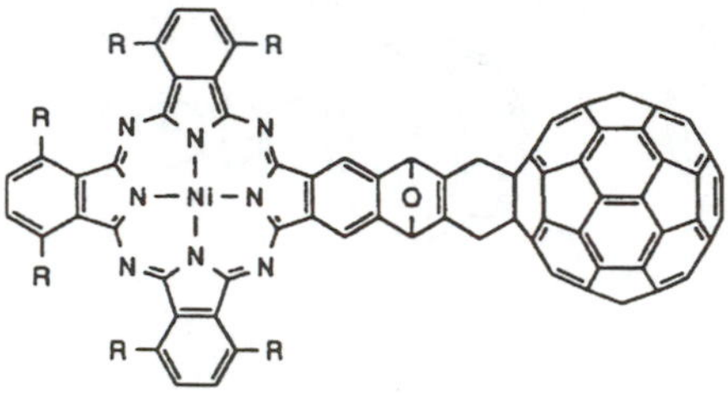

12.35

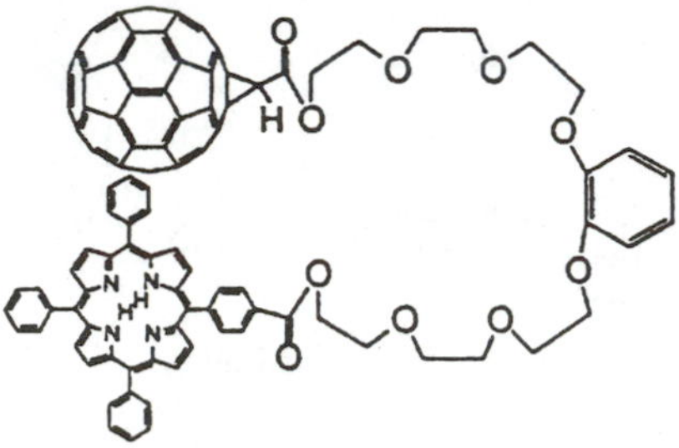

12.36

References

1. A. M. Rao, *et al.*, *Science*, **259** (1993) 955; P. Zhou, Z. Dong, A. M. Rao and P. C. Eklund, *Chem. Phys. Lett.*, **211** (1993) 337.
2. C. N. R. Rao, A. Gindaraj, H. N. Aiyer and R. Seshradi, *J. Phys. Chem.*, **99** (1995) 16814.
3. N. Matsuzawa, M. Ata, D. A. Dixon and G. Fitzgerald, *J. Phys. Chem.*, **98** (1994) 2555.
4. Y. Iwasa *et al.*, *Science*, **264** (1994) 1570.
5. P. Persson, U. Edland, P. Jacobsson, D. Johnels, A. Soldatov and B. Sundqvist, *Chem. Phys. Lett.*, **258** (1996) 540.
6. M. Menon, K. R. Subbaswamy and M. Sawtarie, *Phys. Rev. (B)*, **49** (1994) 13966.
7. G. Wang, K. Komatsu, Y. Murata and M. Shiro, *Nature*, **387** (1997) 583.
8. C. Goze *et al.*, *Phys. Rev. (B)*, **54** (1996) R3676; F. Rachdi, C. Goze and L. Hajji, *J. Phys. Chem. Solids*, **58** (1997) 1645; *Appl Phys.*, **A64** (1997) 295.
9. M. Nuñez-Regueiro *et al.*, *Phys. Rev. Lett.*, **74** (1995) 278.
10. S. Okada and S. Saito, *Phys. Rev. (B)*, **55** (1995) 4039.
11. A. M. Rao *et al.*, *Chem. Phys. Lett.*, **224** (1994) 106.
12. J. M. Hunter, J. L. Fye, N. M. Boivin and M. F. Jarrold, *J. Phys. Chem.*, **98** (1994) 7440.
13. C. Yeretzian, K. Hansen, F. Diederich and R. L. Whetten, *Nature*, **359** (1992) 44.
14. M. Ata, N. Takahashi and K. Nojima, *J. Phys. Chem.*, **98** (1994) 9960.

15. S. W. McElvany, J. H. Callagan, M. M. Ross, L. D. Lamb and D. R. Huffman, *Science*, **260** (1993) 1632.
16. A. Gromov, S. Ballenweg, S. Geisa, S. Lebedkin, W. E. Hull and W. Krätschmer, *Chem. Phys. Lett.*, **267** (1997) 460.
17. R. Taylor, *J. Chem. Soc., Chem. Commun.*, (1994) 1629.
18. E. Albertazzi and F. Zerbetto, *J. Am. Chem. Soc.*, **118** (1996) 2734.
19. G. B. Adams, J. B. Page, M. O'Keefe and O. Sanket, *Chem. Phys. Lett.*, **228** (1994) 460.
20. N. J. Tower and R. Taylor, unpublished work.
21. K. B. Lyons, A. F. Hebard, D. Innis, R. L. Opila, H. L. Carter and R. C. Haddon, *J. Phys. Chem.*, **99** (1995) 16516.
22. Y. Sun, B. Ma, C. E. Bunker and B. Liu, *J. Am. Chem. Soc.*, **117** (1995) 12705.
23. R. Taylor and D. R. M. Walton, *Nature*, **363** (1993) 685.
24. D. A. Loy and R. Assink, *J. Am. Chem. Soc.*, **114** (1992) 3977.
25. B. Z. Tang, S. M. Leung, H. Peng, N. Yu and K. C. Su, *Macromolecules*, **30** (1997) 2848.
26. T. Cao and S. E. Webber, *Macromolecules*, **28** (1995) 3741; C. E. Bunker, G. E. Lawson and Y.-P. Sun, *Macromolecules*, **28** (1995) 3744; A. G. Camp, A. Lary and W. T. Ford, *Macromolecules*, **28** (1995) 7959.
27. S. Mehrotra, A. Nigam and R. Malhotra, *Chem. Commun.*, (1997) 463.
28. H. Nagashima, Y. Saito, M. Kato, T. Kawanashi and K. Itoh, *J. Chem. Soc., Chem. Commun.*, (1992) 377.
29. M. F. Meidine, unpublished work.
30. J. Li *et al.*, *Chem. Lett.*, (1997) 1037.
31. A. Gügel *et al.*, *Tetrahedron*, **52** (1996) 5007.
32. T. Suzuki, Q. Li, K. C. Khemani and F. Wudl, *J. Am. Chem. Soc.*, **114** (1992) 7300.
33. S. Shi, K. C. Khemani, Q. Li and F. Wudl, *J. Am. Chem. Soc.*, **114** (1992) 10656.
34. G. A. Olah *et al.*, *J. Am. Chem. Soc.*, **113** (1991) 9387.
35. D. E. Bergbreiter and H. N. Gray, *J. Chem. Soc., Chem. Commun.*, (1993) 645.
36. L. Dai, A. W. H. Mau, H. J. Greisser, T. H. Spurling and J. W. White, *J. Phys. Chem.*, **99** (1995) 17302.
37. E. T. Samulski *et al.*, *Chem. Mater.*, **4** (1992) 1153.

38. Y. Ederlé and C. Mathis, *Fullerene Sci. & Technol.*, **4** (1996) 1177.
39. C. J. Hawker, *Macromolecules*, **27** (1994) 4836.
40. N. Manalova, I. Rashkov, F. Beguin and H. van Damme, *J. Chem. Soc.*, *Chem. Commun.*, (1993) 3850.
41. K. E. Geckeler and A. Hirsch, *J. Am. Chem. Soc.*, **115** (1993) 3850.
42. A. O. Patil, G. W. Schriver, B. Carstensen and R. D. Lundberg, *Polym. Bull.*, **30** (1993) 187.
43. C. Weis, C. Friedrich, R. Mülhaupt and H. Frey, *Macromolecules*, **28** (1995) 403.
44. Y. Sun, B. Liu and D. K. Moton, *Chem. Commun.*, (1996) 2699.
45. A. Nigam, T. Shekharam, T. Bharadwaj, J. Giovanola, S. Narang and R. Malhotra, *J. Chem. Soc.*, *Chem. Commun.*, (1995) 1547.
46. L. Y. Chiang, L. Y. Wang, S. Tseng, J. Wu and K. Hsieh, *J. Chem. Soc.*, *Chem. Commun.*, (1994) 2675; L. Y. Chiang, L. Y. Wang and C. Kuo, *Macromolecules*, **28** (1995) 7574; S. Tseng, L. Y. Wang, K. Hsieh, W. Liau and L. Y. Chiang, *Fullerene Sci. & Technol.*, **5** (1997) 1313.
47. S. Tseng, L. Y. Wang, K. Hsieh, W. Liau and L. Y. Chiang, *Fullerene Sci. & Technol.*, **5** (1997) 1021.
48. E. Cloutet, Y. Gnanou, J. Fillaut and D. Astruc, *Chem. Commun.*, (1996) 1565.
49. H. L. Anderson, C. Boudon, F. Diederich, J. Gisselbrecht, M. Gross and P. Seiler, *Angew. Chem. Intl. Edn. Engl.*, **33** (1994) 1628.
50. M. Fedurco, D. Costa and A. L. Balch, *Angew. Chem. Intl. Edn. Engl.*, **34** (1995) 194.
51. Y. Wang, *Nature*, **356** (1992) 585; Y. Wang, N. Herron, R. V. Kasowski, A. Suna and K. S. Lee, *The Chemical Physics of Fullerenes 10 (and 5) Years Later*, ed. W. Andreoni, *NATO ASI Series*, **316** (1995) 329.
52. C. J. Hawker, K. L. Wooley and J. M. J. Frechet, *J. Chem. Soc.*, *Chem. Commun.*, (1994) 925.
53. K. L. Wooley *et al.*, *J. Am. Chem. Soc.*, **115** (1993) 9836.
54. X. Camps, H. Schönberger and A. Hirsch, *Chem. Eur.*, **3** (1997) 561.
55. S. Lebedkin *et al.*, *Chem. Phys. Lett.*, **285** (1998) 210.
56. Y. Iwasa, K. Tanoue, T. Mitani, AS. Izuoka, T. Sugawara and T. Yagi, *Chem. Commun.*, (1998) 1411.

57. S. Lebedkin, S. Ballenweg, J. Gross, R. Taylor and W. Krätschmer, *Tetrahedron Lett.*, (1995) 4571; A. B. Smith *et al.*, *J. Am. Chem. Soc.*, **117** (1995) 9359.

58. A. L. Balch, D. A. Costa, W. R. Fawcett and K. Winkler, *J. Phys. Chem.*, **100** (1996) 4823.

59. A. Gromov, S. Lebedkin, S. Ballenweg, A. G. Avent, R. Taylor and W. Krätschmer, *Chem. Commun.*, (1997) 209; R. Taylor, *Proc. Electrochem. Soc.*, **97–14** (1997) 281.

60. P. W. Fowler, D. Mitchell, R. Taylor and G. Seifert, *J. Chem. Soc.*, *Perkin Trans. 2*, (1997) 1901.

61. J. Osterodt and F. Vögtle, *Chem. Commun.*, (1996) 547.

62. T. S. Fabre *et al.*, *J. Org. Chem.*, **63** (1998) 3522.

63. P. Belik, A. Gügel, J. Spikermann and K. Müllen, *Angew. Chem. Intl. Edn. Engl.*, **32** (1993) 78.

64. L. A. Paquette and R. J. Graham, *J. Org. Chem.*, **60** (1995) 2958.

65. L. A. Paquette and W. E. Trego, *Chem. Commun.*, (1996) 419.

66. A. I. de Lucas, N. Martin, L. Sánchez and C. Seone, *Tetrahedron Lett.*, (1996) 9391.

67. J. M. Lawson *et al.*, *J. Org. Chem.*, **61** (1996) 5032.

68. Y. Sun, T. Drovetskaya, R. D. Bolskar, R. Bau, P. D. W. Boyd and C. A. Reed, *J. Org. Chem.*, **62** (1997) 3642.

69. K. Komatsu, N. Takimoto, Y. Murata, T. S. M. Wan and T. Wong, *Tetrahedron Lett.*, **37** (1996) 6153.

70. P. Timmerman, L. E. Witschel, F. Diederich, C. Boudon, J. Gisselbrecht and M. Gross, *Helv. Chim. Acta*, **79** (1996) 6.

71. P. Timmerman, H. L. Anderson, R. Faust, J. Nierengarten, T. Habicher and F. Diederich, *Tetrahedron*, **52** (1996) 4925.

72. A. L. Balch, *The Chemistry of Fullerenes* (ed. R. Taylor), World Scientific, Singapore, 1995, p. 236.

73. F. Diederich, C. Dietrich-Buchecker, J. Nierengarten and J. Sauvage, *J. Chem. Soc.*, *Chem. Commun.*, (1995) 781.

74. S. I. Khan, A. M. Oliver, M. N. Paddon-Row and Y Rubin, *J. Am. Chem. Soc.*, **115** (1993) 4919; R. M. Williams *et al.*, *J. Org. Chem.*, **61** (1996) 5055; M. G. Ranasinghe, A. M. Oliver, D. F. Rothenfluh, A. Salek and M. N. Paddon-Row, *Tetrahedron Lett.*, **37** (1996) 4797; D. M. Guldi *et al.*, *Molecular Nanostructures (Proc. 1997 Int. Winterschool on*

Electronic Properties of Novel Materials) (1998) 118; D. Armspach, E. C. Constable, F. Diederich, C. E. Housecroft and J. Nierengarten, *Chem. Commun.*, (1996) 2009.

75. T. G. Linssen, K. Dürr, M. Hanack and A. Hirsch, *Chem. Commun.*, (1995) 103.

76. I. G. Safanov, P. S. Baran and D. I. Schuster, *Tetrahedron Lett.*, **47** (1997) 8133.

13

Heterofullerenes

This term describes fullerenes in which one or more of the carbons in the cage are replaced by another element. This area of fullerene chemistry is in its infancy, but will expand rapidly now that ways to produce the parent cages in significant quantities have been discovered.

In aromatic chemistry, replacement of C(H) by B and N lead to the isolation of e.g. pyridine, borazarophenanthrene etc., and these elements likewise can replace carbon in fullerenes to give heterofullerenes. Valency considerations dictate that the products must either be radicals (creating the possibility for dimerisation), have additional hydrogen atoms present, or contain more than one heteroatom. Examples are shown in **13.1–13.3** X variously B, N. In principle, it is possible to have structures in which the heteroatoms are in a 1,4-relationship, but all other possibilities require placement of double bonds in many pentagons. This will make such compounds unstable and unlikely to be isolated.

13.1   13.2   13.3

13.1 Borafullerenes

Compounds containing boron, $C_{59}B$, $C_{58}B_2$, $C_{69}B$, $C_{68}B_2$, detected in the mass spectra of the products obtained by using boron/graphite rods in the arc discharge procedure, appear to be less stable than their carbocage analogues.[1] The mass spectrum indicates that traces of fullerenes containing up to six boron atoms are also present.[1]

Borafullerenes show Lewes acidity, enhanced by the electron withdrawal by the cage. Consequently they add ammonia readily (mass spectrometry) but no other chemistry is known.

13.2 Azafullerenes

These compounds can now be produced relatively easily in multimilligram quantities and are likely to constitute a future major fullerene research area, with most of the reactions described elsewhere in this book being applied to them.

$C_{59}NH^+$ was detected in mass spectrometry of 1,2-epimino[60]fullerene ($C_{60}NH$).[2] Subsequently the aza[60]fulleronium ion has been detected (by mass spectrometry), in the product from heating the *N*-(methoxyethoxymethyl) ketolactam derivative of [60]fullerene (see also Sec. 9.1.2 and Fig. 9.7) with *p*-toluenesulphonic acid in 1,2-dichlorobenzene. However, this is readily reduced under the reaction conditions to the corresponding aza[60]fulleren-2-yl radical, **13.4**, which dimerises to 2,2′-bisaza[60]fullerenyl, **13.5** (Fig. 13.1).[3] Both $C_{59}N^+$ and $C_{69}N^+$ have been detected by mass spectrometry in the product from reacting compounds **9.54** (and the [70]fullerene homologue) with

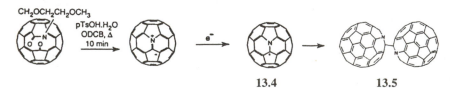

                                                 13.4   **13.5**

Fig. 13.1 Formation of aza[60]fulleren-2-yl, **13.4**, and 2,2′-bisaza[60]fullerenyl, **13.5**, from the *N*(MEM) ketolactam derivative of [60]fullerene.

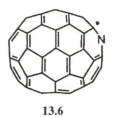

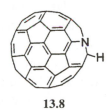

13.6   **13.7**   **13.8**

n-butylamine (which adds to either of the hexagonal rings adjacent to the nitrogens).[4]

If the ketolactam route is employed then it is possible to isolate two isomers of the $C_{69}N^{\bullet}$ radical, *viz.* 1-aza[70]fulleren-2-yl, **13.6** and 2-aza[70]fulleren-1-yl, **13.7**.[5] These readily form dimers, but light can cleave the C-C bond so that ESR spectra can be obtained for the radicals,[6] and likewise for $C_{59}N^{\bullet}$, **13.4**, derived from the dimer precursor, **13.5**.[6,7] The hyperfine coupling constants are 3.6 G and 4.74 G for **13.4** and **13.6**, respectively, the larger value for the latter being attributed to greater localisation of the radical and thus greater overlap with the nitrogen lone pair.[6] However, for **13.7** two hyperfine splittings are found, one of 4.74 G and one of only 0.49 G. The latter value indicates delocalisation of the radical to a site more remote from the nitrogen, location in the equatorial region being suggested. However **13.6** could be expected to undergo this more readily, so some other cause may be responsible.

The $(C_{69}N)_2$ dimer can also be produced by reaction of the [70]fullerene analogue of **9.54** with butylamine and *p*-toluenesulphonic acid and in this case the dimer precursor radical is **13.6** only.[8]

The (green) parent azahydro[60]fullerene $C_{59}NH$, **13.8**, is produced by reacting either the aza[60]fulleren-2-yl radical with hydroquinone, or by cleaving the dimer, **13.5**, with tributyltin hydride.[9] The retention times of $C_{59}NH$ and $(C_{59}N)_2$ on the standard Cosmosil column widely used by fullerene researchers are 7.4 and 14.1 min respectively (toluene eluent, 1 ml/min), *cf. ca.* 8.5 min for C_{60}, so the retention times of derivatives will probably likewise be similarly related to those of the carbocage equivalents. The acidic hydrogen should be readily removed by base to give the aza[60]fulleren-1-ide anion, and so give a more extensive chemistry than for [60]fullerene itself. The first reduction potentials for C_{60}, $C_{59}NH$ and $(C_{59}N)_2$ are −1123 mV, −1106 mV and −992 mV so that electron addition becomes easier along this series;[9] NH has a -I inductive effect relative to the carbon replaced, and this is increased on replacing the hydrogen by a second electron-withdrawing aza[60]fullerenyl group.

Some chemistry of these compounds has commenced. For example, $C_{59}N$ reacts with ICl to give the pentachloro derivative in which the chlorines are disposed radially around the pentagon containing the nitrogen, in a manner analogous to that in $C_{60}Cl_6$ (Fig. 7.7(a)).[10] In the case of this (2,5,10,21, 24-pentachloro-2,5,10,21,24-pentahydroaza[60]fullerene) the extra electron provided by the nitrogen makes addition of an sixth chlorine unnecessary. Moreover, the central pentagon has 6 π electrons and is therefore aromatic, so bestowing aromaticity and stability on the molecule. Creation of this aromatic ring provides a driving force that is likely to dominate the chemistry of these compounds. For example, reduction of $C_{59}N$ by Zn/HCl (which with [60]fullerene gives $C_{60}H_{36}$, Sec. 4.1.3), produces $C_{59}NH_5$ consistent with a structure having the hydrogens arranged radially around the pentagon containing the nitrogen.[11] On the other hand, 33 fluorines are added in fluorination.[12] Notable, here an odd number of fluorines are added in contrast to fluorination of the carbocage equivalent where the number added is always even. This is because one extra fluorine is involved in cleaving the dimer bond.

The alkoxy derivatives $ROC_{59}N$ and $ROC_{69}N$ (R = CH_2CH_2OMe) are obtained as byproducts in the formation of $C_{59}N$ and $C_{69}N$ from **9.54** and the [70]fullerene homologue, the OR group being adjacent to the nitrogen[8] A wider range of aromatic derivatives (with the aromatic attached to the carbon adjacent to the nitrogen, **13.9**) can be obtained in high yield simply by heating $(C_{59}N)_2$ in 1,2-dichlorobenzene with aromatics reactive towards electrophilic aromatic substitution, in the presence of air and *p*-toluenesulphonic acid.[13] All of the derivatives are green and the mechanism is believed to involve oxidative

13.9 (Ar = 4-methoxyphenyl, 4-toluyl, X-1-chloronaphthyl)

13.10

cleavage of the dimer to give the electrophile $C_{59}N^+$. This reaction has the potential for the preparation of a very large range of derivatives.

If the reaction is carried out in the absence of either oxygen or the acid, and with diphenylmethane which has a readily abstractable hydrogen, then instead of electrophilic substitution taking place, radical coupling occurs between $C_{59}N^\bullet$ and $Ph_2CH^\bullet$ occurs to give **13.10**.[14] The potential of the first reduction wave for this is -1082 mV i.e. less than for $C_{59}NH$ due to the electron-withdrawing effect of the Ph_2CH substituent.

This reaction also has the potential for the formation of a wide range of interesting derivatives.

References

1. T. Guo, C. Jin and R. E. Smalley, *J. Phys. Chem.*, **95** (1991) 4948; Y. Chai *et al.*, *J. Phys. Chem.*, **95** (1991) 7564; T. Kimura, T. Sugai and H. Shinohara, *Chem. Phys. Lett.*, **256** (1996) 269; B. Cao, X. Zhou, Z. Shi, Z. Gu, H. Xiao and J. Wang, *Fullerene Sci. & Technol.*, **6** (1998) 639.
2. J. Averdung, H. Luftmann, I. Schlachter and J. Mattay, *Tetrahedron*, **51** (1995) 6977.
3. J. C. Hummelen, B. Knight, J. Pavlovich, R. González and F. Wudl, *Science*, **269** (1995) 1554.
4. I. Lamparth, B. Nüber, G. Schick, A. Skiebe, T. Grösser and A. Hirsch, *Angew. Chem. Intl. Edn. Engl.*, **34** (1995) 2257.
5. C. Bellavia-Lund and F. Wudl, *J. Am. Chem. Soc.*, **117** (1997) 943.

6. K. Hasharoni, C. Bellavia-Lund, M. Kershavarz-K, G. Srdanov and F. Wudl, *J. Am. Chem. Soc.*, **117** (1997) 11128.

7. A. Gruss, K. Dinse, A. Hirsch, B. Nüber and U. Reuther, *J. Am. Chem. Soc.*, **117** (1997) 8728.

8. B. Nüber and A. Hirsch, *Chem. Commun.*, (1996) 1421.

9. M. Keshavarz-K *et al.*, *Nature*, **383** (1996) 147.

10. Personal communication from A. Hirsch.

11. A. D. Darwish, A. Hirsch and R. Taylor, unpublished work.

12. O. V. Boltalina, A. Hirsch and R. Taylor, unpublished work.

13. B. Nüber and A. Hirsch, *Chem. Commun.*, (1998) 405.

14. C. Ballavia-Lund, R. González, J. C. Hummelen, R. G. Hicks, A. Sastre, and F. Wudl, *J. Am. Chem. Soc.*, **119** (1997) 2946.

14

The Chemistry of *Incar*-fullerenes (Endohedral Fullerenes)

14.1 Nomenclature

Fullerenes can have elements trapped within the cages and these compounds have been referred to hitherto as endohedral fullerenes, the formulae being written as e.g. La@C_{82} to indicate that lanthanum is inside the cage. The name is however unfortunate since it confuses with the conventional description of endohedral and exohedral organic compounds, which will be a problem when, in due course, derivatives of cycloaddition are produced. Compounds containing elements within structures have been around long before fullerenes were discovered, and are known as incarceranes. The IUPAC description of the fullerene species is therefore *incar*-fullerenes and the formula for the compound [82]fullerene-*incar*-lanthanum is written as $iLaC_{82}$.

14.2 Fullerenes with Incarcerated Metals

A very large number of these compounds have been detected spectroscopically, and a few have been isolated in macroscopic quantities. The isolation has hitherto been extremely time consuming, so that they have thus far been mainly of theoretical interest. However, methods for formation and isolation of these compounds are continually improving to the extent that chemical studies have now begun. A brief description of their chemistry, and associated properties are described in this chapter.

Metal *incar*-fullerenes so far either detected or isolated are shown in Tables 14.1–14.4.

Table 14.1 Fullerenes encapsulating one atom.

Fullerene	Metal	Ref.
C_{28}	Hf, Ti, U, Zr	1
C_{36}	U	1
C_{44}	K, La, U	1–4
C_{48}	Cs	4
C_{50}	U	1
C_{60}	Li, K, Ca, Co, Y, Cs, Ba, Rb, La, Ce, Pr, Nd, Sm, Eu, Gd, Tb, Dy, Ho, Er, Lu, U	1–3, 5–14
C_{70}	Li, Ca, Y, Ba, La, Ce, Gd, Lu, U	1, 3, 12–14
C_{72}	U	1
C_{74}	Sc, La, Gd, Lu	3, 13, 15
C_{76}	La	2, 15
C_{80}	Ca, Sr, Ba	16
C_{82}	Ca, Sc, Sr, Ba, Y, La, Ce, Pr, Nd, Sm, Eu, Gd, Er, Tm, Lu	3, 13–15,17–41
C_{84}	Ca, Sc, Sr, Ba, La	15, 38, 40

Table 14.2 Fullerenes encapsulating two atoms.

Fullerene	Metal	Ref.
C_{28}	U_2	1
C_{56}	U_2	1
C_{60}	Y_2, La_2, U_2	1, 2, 42
C_{74}	Sc_2	15
C_{76}	La_2,	15
C_{80}	La_2, Ce_2, Pr_2	10, 37, 42–45
C_{82}	Sc_2, Y_2, La_2	15, 21–23, 25, 42, 46
C_{84}	Sc_2, La_2	15, 18, 23, 25, 47–49

Table 14.3 Fullerenes encapsulating three atoms.

Fullerene	Metal	Ref.
C_{82}	Sc_3	24, 25, 50
C_{84}	Sc_3	18, 21, 23, 25, 47

Table 14.4 Fullerenes encapsulating four atoms.

Fullerene	Metal	Ref.
C_{82}	Sc_4	51

14.2.1 *Properties*

The very low yield formation of these compounds, and the sensitivity of many of them to air, mean that their isolation is a very time-consuming process, as noted in the introduction. Nevertheless whereas single mg quantities of them were first obtained around 1992,[3,47] preparation of substantial quantities of iLaC$_{82}$ (27.4 mg) and iGdC$_{82}$ (12.0 mg) have now been described,[39] and these compare with the quantities of C_{60} and C_{70} available in 1990. So, assuming a parallel development, one could anticipate having gram quantities of these materials available in five or so years time. Work has been concentrated so far on derivatives of C_{74} and C_{82} because these tend to be more stable, and hence are formed in higher yields than those obtained from other fullerenes.

The incarcerated metal transfers electrons to the cage thereby altering the properties of the latter. *Incar*-fullerenes can thus be regarded as a 'superatom' in having a positively charged core and a negatively charged cage.[52] For example, in the case of iLaC$_{82}$, three electrons are transferred, and the electronic structure of can thus be represented as iLa^{3+}C$_{82}{}^{3-}$. In general the +3 oxidation level will tend to apply to Sc, Y, La, Ce, Pr, Nd, Gd, Tb, Ho, Er, Lu, and the +2 oxidation level to Ca, Sr, Sm, Eu, Tm, and Yb.[13] One very evident property change is that the *incar*-fullerenes have longer HPLC retention times than their empty-cage analogues (due probably to stronger coordination of the more electron-rich cage with the stationary phase). Moreover, iGdC$_{82}$ has a longer retention time than does iLaC$_{82}$,[39] and it remains to be seen if (for a given cage structure) this trend for the retention time to increase with the atomic mass of the incarcerated atom, will be a general feature. If an incarcerated element can exhibit two different oxidation states (e.g. Sm, Eu, Tm, Yb) then the *incar*-fullerenes in which the element is in the 2+ oxidation state elute with shorter retention times than those in which the element is in the 3+ oxidation state.[51] This follows from the greater charge transfer to the cage in the latter compounds, making them more polar.

Due to the cages being more electron-rich than those of empty fullerenes, they are less soluble in non-polar solvents, but conversely can be selectively separated by polar solvents such as aniline, pyridine, and dimethyl-formamide.[5,41] The band gaps are also of the order of 0.2 eV compared to 1.6 eV for [60]fullerene, and the substantially reduced stabilities rules out characterisation of structures by EI mass spectrosopy since fragmentation occurs, and limits studies to the (considerably fewer) laboratories that are equipped with soft-ionisation techniques. The development of the chemistry of the *incar*-fullerenes can thus be expected to be slower than was the case for the empty fullerenes.

If an odd number of electrons become transferred from the metal to the cage, then an ESR signal is obtained.[15,53,54] For example, in iLaC$_{82}$ the spin of ^{139}La is 7/2 and so by the $n + 1$ rule this gives an octet in the EPR hyperfine spectrum. By contrast iLa$_2$C$_{80}$, iSc$_2$C$_{82}$ and iSc$_2$C$_{84}$ do not give ESR spectra because the nuclear spins for each of the metal ions are paired. On the other hand, trimetal species such as iSc$_3$C$_{82}$ have three interacting Sc nuclei so the maximum nucler spin is $3 \times 7/2 = 21/2$ hence a 22 line spectrum is obtained.[24,25] The electrons transferred to the cage must be in anti-bonding orbitals and thus in principle readily available for conduction. ESR spectra indicate that there are two isomers of iLaC$_{82}$, having 1.159 G and 0.836 G ^{139}La hyperfine splittings,[15,29,35] but the component giving the latter is much more sensitive to air and shows other reactivity differences, attributed to the lanthanum being attached to the outside of the cage.[35,55]

Isomers of *incar*-fullerenes have been isolated by HPLC. Examples are the four isomers of iCaC$_{82}$[38] and three isomers of iTmC$_{82}$ (each is very air stable, but they have quite different electronic properties).[36] Three isomers of iSc$_2$C$_{84}$ have been separated similarly, and ^{13}C NMR shows the main one to be the

$[D_{2d}(II)]$ isomer, the symmetry of the spectrum showing that the positions of the scandium atoms are equivalent within the NMR timescale; the other isomers are $C_s(II)$ and $C_{2v}(III)$ [49] The equivalence of the scandium atoms is confirmed by ^{45}Sc NMR which shows only one line.[56] In both $iLaC_{82}$ and iYC_{82} the metal atoms lie off-centre,[55] so these molecules should be dipolar; for example, $iY^{3+}C_{82}^{3-}$ (the size of which has been determined as 11.5×13.2 Å) is estimated to have a dipole moment of 4 Debye.[51]

Initial electrochemistry studies show for example that $iLaC_{82}$ has five reduction waves and two oxidation waves (the second of which is not reversible) whilst iYC_{82} has four reduction and two oxidation waves.[28,29] Thus these compounds appear to be more readily oxidised than open fullerenes, and this is consistent with the higher electron density on the cage, thus making removal of electron easier. Electrochemical studies also indicate the iLa_2C_{80} is a stronger electron acceptor than monometallofullerenes such as $iLaC_{82}$.[43]

14.2.2 Chemistry

The chemistry of the *incar*-fullerenes, still barely developed, will be governed by the electronic properties arising from the presence of the incarcerated metal. One curious feature is that although $iLaC_{82}$ is a better electron donor than the empty cage (to be expected since the cage is effectively in a 3^- state), it is also a better *acceptor*,[26] and this presumably arises from the presence of the positively charged 'core' in the fullerene. iLa_2C_{80} is a better electron acceptor than $iLaC_{82}$ and some calculations indicate that the first two electrons accepted actually go to the metal, not the cage,[43] though others disagree.[55] The greater electron acceptor properties of the endohedral fullerenes relative to their open-cage counterparts is also indicated by the electron affinity of $iCaC_{60}$ of 3.0 eV, *cf.* 2.65 eV for C_{60} itself.[1,5] The electron affinities of a range of fullerenes containing gadolinium are each *ca.* 0.15 eV greater than their empty-cage counterparts.[57]

$iLaC_{82}$ reacts with diphenyldiazomethane to give mainly a mono adduct (which could be either a methanofullerene or a homofullerene; *cf.* Sec. 9.1), together with traces of bis and tris adducts; the ESR spectrum indicates that five different components (including probably some monoadduct isomers) are produced.[30]

The photochemical reaction of iLaC$_{82}$ with a disilirane [(Ar$_2$Si)$_2$CH$_2$] gives a product which ESR indicates to consist of two isomers, of structures as yet unknown.[32] Reaction also takes place under thermal conditions, whereas empty fullerenes only react with disiliranes under photochemical conditions. iGdC$_{82}$ also reacts similarly with disilirane,[33] as does iLaC$_{82}$ with a digermirane [(Ar$_2$Ge)$_2$CH$_2$], the ESR spectrum from the latter reaction indicating the formation of three regioisomers.[34] In the reactions of disilirane with iLa$_2$C$_{80}$ and iSc$_2$C$_{84}$, addition with the latter fullerene occurs under photochemical conditions only.[55]

14.3 Fullerenes with Incarcerated Nitrogen

The only compound of this type obtained so far is iNC$_{60}$, prepared by heating [60]fullerene to *ca.* 450°C in a glow discharge reactor containing nitrogen. In the recovered fullerene, the ratio of nitrogen-containing- to empty molecules is approximately 10^{-5} to 10^{-6}.[58] The nitrogen, although intrinsically very reactive, does not bond to the inner surface of the cage, because the orbital coefficients on the inside of the cage are small, and also because such bonding would result in increased strain: any carbon involved in bonding would have to move towards the cage centre, increasing the strain on the three other carbon atoms to which it is attached. The absence either of bonding or of charge transfer) is shown by the three-fold degenerate EPR spectrum which is however altered slightly by the presence of addends on the cage. This arises because the distortion of the cage causes the three p-orbitals of the nitrogen to be no longer degenerate. The nitrogen-containing and empty molecules show no differences in reactivity, which is also consistent with the lack of bonding between the nitrogen and the cage (which has been described as a 'chemical Faraday cage').[58]

14.4 Fullerenes Having Incarcerated Noble Gases

Each of helium, neon, argon, krypton and xenon can be incarcerated into fullerenes though the application of high temperature (620°C) and pressure (*ca.* 40,000 psi). The incorporation fractions for [60]fullerene are approximately: He, 0.1%; Ne, 0.2%; Ar, 0.3%, Kr, 0.3% Xe, 0.008%, those for [70]fullerene being similar.[59] The value for helium is about the same as that for fullerenes obtained by the arc-discharge procedure. The noble gases are released from the fullerenes by heating to $1000°C$[60] which is equivalent to *ca.* 80 kcal mol^{-1}. How the noble gas gets inside the cage during the pressurisation process is very unclear, since the energy required even for helium to pass through a hexagon is calculated to be a much higher *ca.* $200 \text{ kcal mol}^{-1}$. The cages must somehow be ruptured and then reform, but this latter step is evidently not always successful since about 50% of the fullerene is lost during the incorporation procedure. Helium can also be incorporated into the radical cations of both [60]- and [70]fullerenes, by passing molecular beams of the radical cations into the stationary noble gas.[61]

The interest in this work from a chemical viewpoint stems from incarceration of the ^3He isotope. Tritium has a half-life of 12.3 years and decays to ^3He, consequently reprocessing of thermonuclear warheads makes substantial quantities of this isotope available at moderate cost. The presence of ^3He can be monitored by NMR, and the spectrum is a probe for the magnetic shielding environment inside the fullerene cavity, in turn reflecting ring currents and hence the aromaticity of the fullerene. Thus, the more aromatic a fullerene, the more upfield should be the signal, hence [70]fullerene (-28.8 ppm) is indicated to be more aromatic than [60]fullerene (-6.3 ppm).[62]

The particular advantages of this analytical technique are two-fold: both each fullerene and each derivative gives a single signal. For the parent fullerenes the peaks are shown in Table 14.5; the assignments for the isomers of [78]- and [84]fullerenes are provisional.[63] Notably, these data indicate that five isomers of [78]fullerene (predicted to be stable) do indeed exist, even though only three have so far been isolated.[64] Moreover, although the data for [84]fullerene is not unambiguous because of the possibility that some of the peaks are due to the oxides of [60]- and [70]fullerenes, nevertheless, the number of peaks is in close agreement with the number of isomers indicated to exist by ^{13}C NMR studies.[65]

For derivatives, the signals appear more upfield for those derived from [60]fullerene (i.e. the derivatives are more aromatic), but more downfield for those derived from [70]fullerene (i.e. the derivatives here are less aromatic), the shifts being of the order of 1–3 ppm for monoaddition.[66] Bis-addition to [60]fullerene produces further shifts in these directions (*ca.* 5 ppm overall), and there is a partial correlation between the shifts and structure, e.g. it increases

Table 14.5 ^{3}He NMR chemical shifts for fullerenes, referenced to ^{3}He gas at 0 ppm.

Chemical shift/ppm	Fullerene	Chemical shift/ppm	Fullerene
−6.3	[60]	−11.1	[84-?]
−28.8	[70]	−10.5	[84-?]
−18.72	[76-D_2]	−9.65	[84-?]
−11.925	[78-D_3]	−9.61	[84-?]
−16.90 (main)	[78-C_{2v}(I)]	−8.96 (main)	[84-D_2]
−16.78	[78-C_{2v}(II)]	−8.41	[84-?]
−17.59	[78-?]	−8.37	[84-?]
−18.58	[78-?]	−7.55	[84-?]
		−7.51	[84-?]

in the series *cis*-3 < *cis*-1 < *cis*-2 ≈ *trans*-4.[67] However, further addition up to the hexakis level produces only trivial changes. By contrast, polyaddition to [70]fullerene (up to the tetra-kis level) appears to produce regular increases in shifts.[68]

The incarceration of ^{3}He has been used to show that two isomers of $C_{60}H_{36}$ are formed, and in a ratio of *ca.* 3:1.[69] This parallels the formation in a similar ratio of two isomers of $C_{60}F_{36}$ which have been fully characterised by ^{19}F NMR as having T and C_3 symmetry see Sec. 7.1.1.2). Moreover, the ^{3}He NMR spectrum of $i^3HeC_{60}F_{36}$ also consists of two peaks in a similar ratio to the above and in the same region of the spectrum, providing further evidence that the same isostructural isomers are formed in both hydrogenation and fluorination.[70]

References

1. T. Guo *et al.*, *Science*, **257** (1992) 1661.
2. J. R. Heath *et al.*, *J. Am. Chem. Soc.*, **107** (1985) 7779.

3. Y. Chai *et al.*, *J. Phys. Chem.*, **95** (1991) 7564.
4. F. D. Weiss, S. C. O'Brien, J. L. Eklund, R. F. Curl and R. E. Smalley, *J. Am. Chem. Soc.*, **110** (1988) 4464.
5. L. S. Wang *et al.*, *Chem. Phys. Lett.*, **207** (1993) 354.
6. L. S. Wang, J. M. Alford, Y. Chai, M. Diener and R. E. Smalley, *Z. Phys. D*, **26** (1993) 297.
7. D. S. Bethune *et al.*, *Nature*, **363** (1993) 605.
8. R. F. Curl, *Carbon*, **30** (1992) 1149.
9. R. Huang, H. Li, W. Lu and S. Yang, *Chem. Phys. Lett.*, **228** (1994) 111.
10. E. G. Gillan, C. Yeretzian, K. S. Min, M. M. Alvarez, R. L. Whetten and R. B. Kaner, *J. Phys. Chem.*, **96** (1992) 6869.
11. Y. Kubozono *et al.*, *Chem. Lett.*, (1995) 457.
12. Y. Kubozono *et al.*, *J. Am. Chem. Soc.*, **118** (1996) 6998.
13. L. Moro, R. S. Ruoff, C. H. Becker, D. C. Lorents and R. Malhotra, *J. Phys. Chem.*, **97** (1993) 6801.
14. A.Gromov, W. Krätschmer, N. Krawez, R. Tellgmann and E. E. B. Campbell, *Chem. Commun.*, (1997) 2003.
15. H. Shinohara *et al.*, *Mat. Sci. and Eng.*, **B19** (1993) 25; *J. Phys. Chem.*, **97** (1993) 4259.
16. T. J. S. Dennis and H. Shinohara, *Chem. Commun.*, (1998) 883.
17. M. E. J. Boonman *et al.*, *Physica B.*, **211** (1995) 323.
18. R. D. Johnson, M. S. de Vries, J. Salem, D. S. Bethune and C. S. Yannoni, *Nature*, **355** (1992) 239.
19. K. Kikuchi *et al.*, *Chem. Phys. Lett.*, **216** (1993) 67.
20. Y. Achiba, T. Wakabayashi, T. Moriwaki, S. Suzuki and H. Shiromaru, *Mat. Sci. and Eng.*, **B19** (1993) 14.
21. A. Bartl, L. Dunsch, J. Froehner and U. Kirbach, *Chem. Phys. Lett.*, **229** (1994) 115.
22. J. H. Weaver *et al.*, *Chem. Phys. Lett.*, **190** (1992) 460.
23. H. Shinohara, H. Sato, Y. Saito, M. Ohkohchi and Y. Ando, *J. Phys. Chem.*, **96** (1992) 3571.
24. H. Shinohara *et al.*, *Nature*, **357** (1992) 52.
25. C. S. Yannoni *et al.*, *Science*, **256** (1992) 1191.
26. T. Suzuki, Y. Maruyama, T. Kato, K. Kikuchi and Y. Achiba, *J. Am. Chem. Soc.*, **115** (1993) 11006.

27. T. Kato, S. Suzuki, K. Kikuchi and Y. Achiba, *J. Phys. Chem.*, **97** (1993) 13425.

28. K. Kikuchi, Y. Nakao, S. Suzuki, Y. Achiba, T. Suzuki and Y. Maruyama, *J. Am. Chem. Soc.*, **116** (1994) 9367.

29. K. Yamamoto, H. Funasaka, T. Takahashi, T. Akasaka, T. Suzuki and Y. Maruyama, *J. Phys. Chem.*, **98** (1994) 12831.

30. T. Suzuki *et al.*, *J. Am. Chem. Soc.*, **117** (1995) 9606.

31. H. Shinohara, M. Inakuma, M. Kishida, S. Yamazaki, T. Hashizume and T. Sakurai, *J. Phys. Chem.*, **99** (1995) 13769.

32. T. Akasaka *et al.*, *Nature*, **374** (1995) 600.

33. T. Akasaka *et al.*, *J. Chem. Soc.*, *Chem. Commun.*, (1995) 1343.

34. T. Akasaka *et al.*, *Tetrahedron*, **52** (1996) 5015.

35. Y. Saito, S. Yokoyama, M. Inakuma and H. Shinohara, *Chem. Phys. Lett.*, **250** (1996) 80.

36. U. Kirbach and L. Dunsch, *Angew. Chem. Intl. Edn. Engl.*, **35** (1996) 2380.

37. J. Ding and S. Yang, *J. Am. Chem. Soc.*, **118** (1996) 11254.

38. Z. Xu, T. Nakane and H. Shinohara, *J. Am. Chem. Soc.*, **118** (1996) 11309.

39. H. Funasaka, K. Sugiyama, K. Yamamoto and T. Takahashi, *J. Phys. Chem.*, **99** (1995) 1826.

40. T. J. S. Dennis and H. Shinohara, *Chem. Phys. Lett.*, **278** (1997) 107.

41. J. Ding, N. Liu, L. Weng, N. Cue and S. Yang, *Chem. Phys. Lett.*, **261** (1996) 92; J. Ding, L. Weng and S. Yang, *J. Phys. Chem.*, **100** (1996) 11120.

42. R. E. Smalley, *Acc. Chem. Res.*, **25** (1992) 98.

43. T. Suzuki *et al.*, *Angew. Chem. Intl. Edn. Engl.*, **34** (1995) 1094; T. Akasaka *et al.*, *Angew. Chem. Intl. Edn. Engl.*, **36** (1997) 1643.

44. M. M. Alvarez, E. G. Gillan, K. Holczer, R. B. Kaner, K. S. Min and R. L. Whetten, *J. Phys. Chem.*, **95** (1991) 10561.

45. M. M. Ross, H. J. Nelson, J. H. Callahan and S. W. McElvany, *J. Phys. Chem.*, **96** (1992) 5231.

46. H. Shinohara, H. Sata, Y. Saito, M. Ohkohchi and Y. Ando, *Rapid Commun. Mass Spectrom.*, **6** (1992) 413.

47. R. D. Johnson, D. S. Bethune and C. S. Yannoni, *Acc. Chem. Res.*, **25** (1992) 169.

48. R. Beyers *et al.*, *Nature*, **370** (1994) 196.
49. E. Yamamoto, M. Tansho, T. Tomiyama, H. Shinohara, H. Kawahara and Y. Kobayashi, *J. Am. Chem. Soc.*, **118** (1996) 2293.
50. T. Kato, S. Bandou, M. Inakuma and H. Shinohara, *J. Phys. Chem.*, **99** (1995) 856.
51. H. Shinohara, reported at the Electrochemical Society Meeting, Montreal, Mat 1997.
52. S. Nagase and K. Kobayashi, *J. Chem. Soc., Chem. Commun.*, (1994) 1837.
53. S. Suzuki *et al.*, *J. Phys. Chem.*, **96** (1992) 7159.
54. M. Hoinkis *et al.*, *Chem. Phys. Lett.*, **198** (1992) 461.
55. S. Nagase, K. Kobayashi and T. Akasaka, *Bull. Chem. Soc. Jpn.*, **69** (1996) 2131.
56. Y. Miyake *et al.*, *J. Phys. Chem.*, in press.
57. O. V. Boltalina, I. N. Ioffe, I. G. Sorokin and L. N. Sidorov, *J. Phys. Chem.*, **101** (1997) 9561.
58. B. Pietzak *et al.*, *Chem/. Phys. Lett.*, **279** (1997) 259.
59. M. Saunders *et al.*, *J. Am. Chem. Soc.*, **116** (1994) 2193; M. Saunders, H. A. Jiménez-Vázquez, R. J. Cross and R. J. Poreda, *Science*, **259** (1993) 1428.
60. R. Shimshi, A. Khong, H. A. Jiménez-Vázquez, R. J. Cross and M. Saunders, *Tetrahedron*, **52** (1996) 5143.
61. J. H. Callahan, M. M. Ross, T. Weiske and H. Schwarz, *J. Phys. Chem.*, **97** (1993) 20, and references contained therein.
62. M. Saunders, H. A. Jiménez-Vázquez, R. J. Cross, S. Mroczkowski, D. I. Freedberg and F. A. L. Anet, *Nature*, **367** (1994) 256.
63. M. Saunders *et al.*, *J. Am. Chem. Soc.*, **117** (1995) 9305.
64. K. Kikuchi *et al.*, *Nature*, **357** (1992) 142; R. Taylor, *J. Chem. Soc., Perkin Trans. 2*, (1992) 3; R. Taylor *et al.*, *J. Chem. Soc., Chem. Commun.*, (1992) 1403; *J. Chem. Soc., Perkin Trans. 2*, (1993) 1029.
65. A. G. Avent, D. Dubois, A. Penicaud and R. Taylor, *J. Chem. Soc., Perkin Trans. 2*, (1997) 1907; T. J. S. Dennis and H. Shinohara, personal communication.
66. A. B. Smith *et al.*, *J. Am. Chem. Soc.*, **116** (1994) 10831; M. Saunders, R. J. Cross, H. A. Jiménez-Vázquez, R. Shimshi and A. Khong, *Science*, **271** (1996) 1693.

67. R. J. Cross *et al.*, *J. Am. Chem. Soc.*, **118** (1996) 11454.
68. M. Rüttiman *et al.*, *Chem. Eur. J.*, **3** (1997) 1071.
69. W. E. Billups *et al.*, *Tetrahedron Lett.*, **38** (1997) 175, 179.
70. O. V. Boltalina, A. Khong, M. Saunders and R. Taylor, unpublished work.